Convenient Solutions
to an Inconvenient Truth

ENVIRONMENT AND DEVELOPMENT

A fundamental element of sustainable development is environmental sustainability. Hence, this series was created in 2007 to cover current and emerging issues in order to promote debate and broaden the understanding of environmental challenges as integral to achieving equitable and sustained economic growth. The series will draw on analysis and practical experience from across the World Bank and from client countries. The manuscripts chosen for publication will be central to the implementation of the World Bank's Environment Strategy, and relevant to the development community, policymakers, and academia. Topics addressed in this series will include environmental health, natural resources management, strategic environmental assessment, policy instruments, and environmental institutions, among others.

Titles in this series:

Convenient Solutions to an Inconvenient Truth: Ecosystem-Based Approaches to Climate Change

Environmental Flows in Water Resources Policies, Plans, and Projects: Findings and Recommendations

Environmental Health and Child Survival: Epidemiology, Economics, and Experiences

International Trade and Climate Change: Economic, Legal, and Institutional Perspectives

Poverty and the Environment: Understanding Linkages at the Household Level

Strategic Environmental Assessment for Policies: An Instrument for Good Governance

Convenient Solutions to an Inconvenient Truth

Ecosystem-Based Approaches to Climate Change

THE WORLD BANK
Washington, DC

© 2010 The International Bank for Reconstruction
and Development / The World Bank
1818 H Street NW
Washington DC 20433
Telephone: 202-473-1000
Internet: www.worldbank.org
E-mail: feedback@worldbank.org

All rights reserved

1 2 3 4 13 12 11 10

This volume is a product of the staff of the International Bank for Reconstruction and Development / The World Bank. The findings, interpretations, and conclusions expressed in this volume do not necessarily reflect the views of the Executive Directors of The World Bank or the governments they represent.

The World Bank does not guarantee the accuracy of the data included in this work. The boundaries, colors, denominations, and other information shown on any map in this work do not imply any judgement on the part of The World Bank concerning the legal status of any territory or the endorsement or acceptance of such boundaries.

RIGHTS AND PERMISSIONS

The material in this publication is copyrighted. Copying and/or transmitting portions or all of this work without permission may be a violation of applicable law. The International Bank for Reconstruction and Development / The World Bank encourages dissemination of its work and will normally grant permission to reproduce portions of the work promptly.

For permission to photocopy or reprint any part of this work, please send a request with complete information to the Copyright Clearance Center Inc., 222 Rosewood Drive, Danvers, MA 01923, USA; telephone: 978-750-8400; fax: 978-750-4470; Internet: www.copyright.com.

All other queries on rights and licenses, including subsidiary rights, should be addressed to the Office of the Publisher, The World Bank, 1818 H Street NW, Washington, DC 20433, USA; fax: 202-522-2422; e-mail: pubrights@worldbank.org.

ISBN: 978-0-8213-8126-7
eISBN: 978-0-8213-8127-4
DOI: 10.1596/978-0-8213-8126-7

Library of Congress Cataloging-in-Publication Data
Convenient solutions to an inconvenient truth : ecosystem-based approaches to climate change.
  p. cm. — (Environment and development)
  Includes bibliographical references and index.
  ISBN 978-0-8213-8126-7 — ISBN 978-0-8213-8127-4 (electronic)
  1. Climatic changes. 2. Conservation of natural resources. 3. Ecosystem management. I. World Bank.
  QC903.C665 2009
  363.738'74—dc22
                                                              2009040427

Cover photos: Giraffes/Vaclav Volrab/Shutterstock Images LLC;
  Desert/Anthony Whitten/World Bank
Cover design: Auras Design, Silver Spring, Maryland

CONTENTS

ix Acknowledgments
xi Abbreviations and Glossary

1 Overview
2   Ecosystem-Based Mitigation
3   Ecosystem-Based Adaptation

CHAPTER 1
9 The World Bank and Biodiversity Conservation: A Contribution to Action for Climate Change
11 Impacts of Climate Change on Ecosystems and Biodiversity
14 Impacts on Human Communities and Livelihoods
17 Why Protecting Ecosystems and Biodiversity Matters in a Changing World

CHAPTER 2
21 Natural Ecosystems and Mitigation
26 Securing Carbon Stores through Protection and Restoration of Natural Ecosystems
35 Protected Areas: A Convenient Solution to Protect Carbon Sinks and Ecosystem Services
38 Coastal and Marine Systems as Carbon Reservoirs
39 Investing in Alternative Energy

CHAPTER 3
49 Ecosystem-Based Adaptation: Reducing Vulnerability
50 Conserving Biodiversity under Climate Change
52 Maintaining and Restoring Natural Ecosystems
53 Reducing Vulnerability
57 Adopting Indigenous Knowledge to Adapt to Climate Change
57 Adaptation in Coastal Areas
61 Marine Protected Areas
63 Investing in Ecosystems versus Infrastructure

CHAPTER 4
69 Biodiversity Conservation and Food, Water, and Livelihood Security: Emerging Issues
69 Agriculture and Biodiversity
70 Impacts of Climate Change on Agriculture
74 Sustainable Land Management
77 Managing Invasive Alien Species

81 Protecting Natural Ecosystems for Water Services
84 Natural Water Towers

CHAPTER 5
87 Implementing Ecosystem-Based Approaches to Climate Change
88 Looking Forward: The Strategic Framework for Climate Change and Development
89 Growing Forest Partnerships
90 Developing Financing Mechanisms to Support Ecosystem-Based Approaches
91 Climate Investment Funds
91 Reducing Emissions from Deforestation and Degradation
93 Forest Funds

Appendix
97 Securing Carbon Finance at the World Bank: Minimum Project Requirements

101 References

105 Index

Boxes
12 1.1 Monitoring the Impact of Climate Change in a Biodiversity Hot Spot
13 1.2 Climate Change and Biodiversity Loss in Hövsgöl National Park, Mongolia
14 1.3 Likely Regional Impacts on Human Communities and Livelihoods
23 2.1 Reforestation under the BioCarbon Fund
25 2.2 Building Resilience by Promoting Native Vegetation in Mali
29 2.3 Economic Arguments for Sustainable Forest Management
30 2.4 Carbon and Conservation in the Forests of Indonesia
32 2.5 Nariva Wetland Restoration and Carbon Offsets in Trinidad and Tobago
34 2.6 Safeguarding Grasslands to Capture Carbon: Lessons from China
37 2.7 Amazon Region Protected Areas Program: A Storehouse for Carbon and Biodiversity
38 2.8 Crucial Role of Oceans in Climate Change
39 2.9 The Economics for Protecting Coral Reefs
40 2.10 The Manado Ocean Declaration
41 2.11 Nakai Nam Theun: Forest Conservation to Protect a Hydropower Investment in Lao PDR
46 2.12 Biofuels: Too Much of a Good Thing?
51 3.1 Biological Corridors in a Changing World
54 3.2 Restoring the Lower Danube Wetlands
55 3.3 Rebuilding Resilience in Wetland Ecosystems
56 3.4 Ecomarkets in Costa Rica

58	3.5	Measures to Address Climate Change in the Salinas and Aguada Blanca National Reserve in Peru
59	3.6	Investing in Mangroves
60	3.7	Addressing the Impacts of Climate Change on Ocean Ecosystems and Coastal Communities
62	3.8	COREMAP: Coral Reef Rehabilitation and Management Project in Indonesia
64	3.9	Coral Triangle Initiative on Coral Reefs, Fisheries, and Food Security
65	3.10	Protecting Natural Forests for Flood Control
71	4.1	Insects and Orange Juice: Paying for Ecosystem Services in Costa Rica
73	4.2	Water Tanks for Irrigation in Andhra Pradesh, India
74	4.3	Adaptation to Climate Change: Exploiting Agrobiodiversity in the Rain-fed Highlands of the Republic of Yemen
76	4.4	Conservation Farming in Practice in South Africa
78	4.5	Payments for Environmental Services to Protect Biodiversity and Carbon in Agricultural Landscapes
80	4.6	A Cost-Effective Solution for Increasing Water Supply: Removing Invasive Species in South Africa
81	4.7	The Downstream Benefits of Forest Conservation in Madagascar
82	4.8	Lakes in the Central Yangtze River Basin, China
85	4.9	Wastewater Treatment with Wetlands
86	4.10	Protected Areas as Water Towers: Mongolia's Least Costly Solution
94	5.1	Principles for Leveraging Benefits from REDD for the Poor
96	5.2	Can Carbon Markets Save Sumatran Tigers and Elephants?

Figures

10	1.1	Approximate Stores and Fluxes of Carbon
22	2.1	Likely Changes to Earth Systems Depending on Mitigation Activities Undertaken
27	2.2	Forest Area and Forest Carbon Stocks on Lands Suitable for Major Drivers of Tropical Deforestation
35	2.3	Amount of Carbon Stored in Protected Areas, by Region

Tables

16	1.1	Five Climate Threats and the Countries Most at Risk
19	1.2	Total Biodiversity Investments, by Year and Source of Funding
26	2.1	Carbon Stocks in Natural Ecosystems and Croplands
43	2.2	Known Invasive Species Proposed as Suitable for Biofuel Production
66	3.1	Ecosystem-Based Approaches to Defend against Natural Disasters
68	3.2	Exploring the Impacts and Offsets of Infrastructure Projects to Protect Carbon Sinks and Ecosystem Services
92	5.1	Potential Benefits from Ecosystem Protection

Acknowledgments

This book was prepared by a team led by Kathy MacKinnon (task team leader), assisted by Valerie Hickey and Junu Shrestha (Biodiversity Team, Environment Department), with substantial material on adaptation from Walter Vergara (Latin America and the Caribbean Region), and contributions from Marjory-Anne Bromhead, Christophe Crepin, Karsten Ferrugiel, and Gayatri Kanungo (Africa Region); Emilia Battaglini, Maurizio Guadagni, and Karin Shepardson (Europe and Central Asia Region); Joe Leitmann and Tony Whitten (East Asia and Pacific Region); Enos Esikuri, Marea Hatziolos, Claudia Sobrevila, and Klas Sander (Environment Department); Gunars Platais, Adriana Moreira, Stefano Pagiola, Juan Pablo Ruiz, and Jocelyne Albert (Latin America and the Caribbean Region); Kanta Kumari Rigaud (Middle East and North Africa Region); Richard Damania and Malcolm Jansen (South Asia Region); Rafik Hirji (Water Unit of the Energy, Transport, and Water Department); Stephen Ling (Global Facility for Disaster Reduction and Recovery Division); and Sachiko Morita and Charles di Leva (Environmentally and Socially Sustainable Development and International Law). Thanks are due to the Global Environment Facility (GEF) regional coordinators and to Steve Gorman and the GEF Anchor Team for support and advice with regard to the GEF portfolios. The team is also grateful for the helpful comments provided on various versions of the draft publication by peer reviewers Tony Whitten, Adriana Moreira, Ian Noble, and Erick Fernandes as well as Rafik Hirji and Sandy Chang. This publication and other publications about the Bank's work on biodiversity are available online at http://www.worldbank.org/biodiversity.

Abbreviations and Glossary

ACG	Area de Conservación de Guanacaste
AFEP	Aceh Forest and Environment Project
AHTEG	Ad Hoc Technical Expert Group on Biodiversity and Climate Change
ARPA	Amazon Region Protected Areas
CARICOM	Caribbean Community
CBD	Convention on Biological Diversity
CDM	Clean Development Mechanism under the Kyoto Protocol
CEPF	Critical Ecosystem Partnership Fund
CER	Certified emissions reduction, issued in return for a reduction of atmospheric carbon emissions through projects under the Kyoto Protocol's CDM. One CER equals an emission reduction of 1 ton of CO_2. Storing 12 grams of carbon would be equivalent to reducing CO_2 emissions by 44 grams.
CFR	Cape Floristic Region, the smallest floral kingdom, located in South Africa
CIFs	Climate Investment Funds
CO_2	carbon dioxide, a greenhouse gas
CO_2e	carbon dioxide equivalent, which is equivalent to the concentration of CO_2 that would cause the same level of warming as a given type and concentration of another greenhouse gas
COREMAP	Coral Reef Rehabilitation and Management Project
CPACC	Caribbean Planning for Adaptation to Climate Change
CRTR	Coral Reef Targeted Research and Capacity Building for Management Program
EBA	Endemic Bird Area
FCPF	Forest Carbon Partnership Facility
FIP	Forest Investment Program
FONAFIFO	Fondo Nacional de Financiamiento Forestal
FUNBIO	Brazilian Biodiversity Fund
GDP	gross domestic product
GEF	Global Environment Facility
GFP	Growing Forest Partnerships initiative

GHG	greenhouse gas, including carbon dioxide, methane, and nitrous oxide
GISP	Global Invasive Species Programme
GtC	gigaton of carbon
IBA	Important Bird Area
IPCC	Intergovernmental Panel on Climate Change
IUCN	International Union for Conservation of Nature
KfW	Kreditanstalt für Wiederaufbau
LULUCF	Land Use, Land Use Change, and Forestry
MABC	Mesoamerican Biological Corridor
Mt CO_2e	megaton of carbon dioxide equivalent
NGO	nongovernmental organization
PES	payments for ecosystem services
PPCR	Pilot Program for Climate Resilience
PSA	Program of Payment for Environmental Services (Costa Rica)
REDD	Reducing Emissions from Deforestation and Forest Degradation
SCF	Strategic Climate Fund
SFCCD	Strategic Framework for Climate Change and Development
tC	ton of carbon
UNESCO	United Nations Educational, Scientific, and Cultural Organization
UNFCCC	United Nations Framework Convention on Climate Change
UN-REDD	United Nations Collaborative Program on Reducing Emissions from Deforestation and Forest Degradation in Developing Countries
WWF	World Wide Fund for Nature

Note: All dollar amounts are U.S. dollars unless otherwise indicated.

Overview

THE WORLD BANK'S MISSION is to alleviate poverty and support sustainable development. Climate change is a serious environmental challenge that could undermine these goals. Since the Industrial Revolution, the mean surface temperature of Earth has increased an average 0.6° C (Celsius) due to the accumulation of greenhouse gases (GHGs) in the atmosphere. Most of this change has occurred in the last 30 to 40 years, and the rate of increase is accelerating. These rising temperatures will have significant impacts at a global scale and at local and regional levels. While reducing GHG emissions and reversing climate change are important long-term goals, many of the impacts of climate change are already in evidence. As a result, governments, communities, and civil society are increasingly concerned with anticipating the future effects of climate change, while searching for strategies to mitigate, and adapt to, its current and future effects.

Global warming and changes in climate have already had observed impacts on natural ecosystems and species. Natural systems such as wetlands, mangroves, coral reefs, cloud forests, and Arctic and high-latitude ecosystems are especially vulnerable to climate-induced disturbances. However, enhanced protection and management of biological resources and habitats can mitigate the impacts and contribute to solutions as nations and communities strive to adapt to climate change. Biodiversity is the foundation and mainstay of agriculture, forests, and fisheries. Biological resources provide the raw materials for livelihoods, agriculture, medicines, trade, tourism, and industry. Forests, grasslands, freshwater, and marine and other natural ecosystems provide a range of services often not recognized in national economic accounts but vital to human welfare: regulation of water flows and water quality, flood control, pollination, decontamination, carbon sequestration, soil conservation, and nutrient and hydrological cycling.

Current efforts to address climate change focus mainly on reducing GHG emissions by adopting cleaner energy strategies and on reducing the vulnerability of communities at risk by improving infrastructure to meet new energy and water needs. This book offers a compelling argument for including ecosystem-based approaches to mitigation and adaptation as an essential pillar in national strategies to address climate change. Such ecosystem-based strategies can offer cost-effective, proven, and sustainable solutions that contribute to, and complement, other national and regional adaptation strategies.

Ecosystem-Based Mitigation

Terrestrial and oceanic ecosystems play a significant role in the global carbon cycle. Natural habitats are a net store of carbon, with terrestrial ecosystems removing 3 gigatons of carbon (GtC) and oceans another 1.7 GtC from the atmosphere every year. Worldwide, soils alone store an estimated 2,000 GtC. Natural ecosystems serve as major carbon stores and sinks, mitigating and reducing GHG emissions from energy production or land use changes. Biological mitigation of GHGs can occur through (a) sequestration by increasing the size of carbon pools (for example, through afforestation, reforestation, and restoration of natural habitats), (b) maintenance of existing carbon stores (for example, avoiding deforestation or protecting wetlands), (c) maintenance of the ocean carbon sink, and (d) substitution of fossil fuel energy with cleaner technologies based on biomass. The estimated upper limit of the global potential of biological mitigation through afforestation, reforestation, avoided deforestation, and improved agriculture, grazing, and forest management is 100 GtC by 2050, which is equivalent to about 10–20 percent of projected fossil fuel emissions during that period.

Forests cover about 30 percent of total land area, but they store about 50 percent of Earth's terrestrial carbon (1,150 GtC) in plant biomass, litter, debris, or soil. About 20 percent of total GHG emissions are caused by deforestation and land use changes, but in tropical regions emissions attributable to land clearance are much higher, up to 40 percent of national totals. Reducing emissions from deforestation and forest degradation (REDD) is the forest mitigation option with the greatest potential for maintaining carbon stocks in standing forests over the short term.

Various types of wetlands—including swamp forests, mangroves, peatlands, mires, and marshes—are also important carbon sinks and stores. Anaerobic conditions in inundated wetland soils and slow decomposition rates contribute to long-term storage of carbon in the soil and formation of carbon-rich peats.

Peatlands can extend up to 20 meters in depth and represent some 25 percent of the world's soil carbon pool, an estimated 550 GtC; they are estimated to sequester another 0.3 ton of carbon (tC) per hectare per year. Maintaining and restoring wetland habitats protect these carbon sinks, while clearance and drainage can lead to peat collapse and further carbon emissions.

Grasslands occur on every continent except Antarctica and constitute about 34 percent of the global stock of terrestrial carbon. Changes in grassland vegetation due to overgrazing, conversion to cropland, desertification, fire, fragmentation, and introduction of non-native species affect their capacity to store carbon and may, in some cases, even lead to grasslands becoming a net source of carbon dioxide (CO_2). For example, grasslands may lose 20 to 50 percent of their soil organic carbon content through cultivation, soil erosion, and land degradation. Burning of biomass, especially in tropical savannas, contributes more than 40 percent of gross global emissions of carbon dioxide.

Oceans, too, are substantial reservoirs of carbon, holding approximately 50 times more carbon than is held in the atmosphere. They are efficient in taking up atmospheric carbon through plankton photosynthesis, mixing of atmospheric CO_2 with seawater, formation of carbonates and bicarbonates, conversion of inorganic carbon to particulate organic matter, and burial of carbon-rich particles in the deep sea.

Enhanced protection and improved management of natural ecosystems clearly can contribute both to reductions in GHG emissions and to carbon sequestration. Many protected areas, for instance, overlie areas of high carbon stocks. Globally ecosystems within terrestrial protected areas store more than 312 GtC or 15 percent of the terrestrial stock of carbon, although the extent to which these stocks are protected varies with the effectiveness of management.

Ecosystem-Based Adaptation

Adaptation is becoming an increasingly important part of the development agenda. Protecting forests, wetlands, coastal habitats, and other natural ecosystems can provide social, economic, and environmental benefits, both directly through more sustainable management of biological resources and indirectly through protection of ecosystem services. Natural ecosystems maintain the full range of goods and ecosystem services, including natural resources such as water, timber, and fisheries on which human livelihoods depend; these services are especially important to the most vulnerable sectors of society. Protected areas, and the natural habitats within them, can protect watersheds and regulate the flow and quality of water, prevent soil erosion, influence rainfall regimes and local climate, conserve renewable harvestable resources and genetic reservoirs, and protect breeding stocks, natural pollinators, and seed dispersers, which maintain ecosystem health. Over the last decade, more and more Bank projects have been making explicit linkages between conservation and sustainable use of natural ecosystems, carbon sequestration, and watershed values associated with erosion control, clean water supplies, and flood control. Better protection and management of key habitats and natural resources can benefit poor, marginalized, and indigenous communities by protecting ecosystem services and maintaining access to resources even during difficult times, including drought and disaster.

In response to climate change, many countries are likely to invest in even more infrastructure for coastal defenses and flood control to reduce the vulnerability of human settlements to climate change. As water shortages become more frequent and severe, the demand for new irrigation facilities and new reservoirs will grow. Forests, wetlands, or other natural habitats play an important role in protecting high-quality water supplies. Similarly, natural ecosystems can reduce vulnerability to natural hazards and extreme climatic events and complement, or substitute for, more expensive infrastructure investments to protect coastal and riverine settlements. Floodplain forests and coastal mangroves provide storm protection and coastal defenses and serve as safety barriers against natural hazards such as floods, hurricanes, and tsunamis, while wetlands filter pollutants and serve as water recharge areas and as nurseries for local fisheries. Traditional engineered solutions often work against nature, particularly when they aim to constrain regular ecological cycles, such as annual river flooding and coastal erosion, and could further threaten ecosystem services if the construction of dams, seawalls, and flood canals leads to habitat loss. Instead, in Argentina and Ecuador, flood control projects utilize the natural storage and recharge properties of critical forests and wetlands by integrating them into "living with floods" strategies that incorporate forest protected areas and riparian corridors. These simple and effective solutions protect both communities and natural capital.

Three of the world's greatest challenges over the coming decades will be biodiversity loss, climate change, and water shortages. Biodiversity loss will lead to the erosion of ecosystem services and will exacerbate vulnerability to the impacts of climate change. Climate change will lead to water scarcity, increased risk of crop failure, pest infestation, overstocking, permanent degradation of grazing lands, and livestock deaths. Water shortages will affect agricultural productivity, food security, and human health. Impacts from these challenges are already imposing severe economic and social costs, and they are likely to become more severe as climate change continues, particularly affecting vulnerable communities.

Changing climate and rainfall patterns are expected to have significant impacts on agricultural productivity, especially in arid and semiarid regions that are marginal for agriculture. Most climate modeling scenarios indicate that the drylands of West and Central Asia and North Africa, for instance, will be severely affected by droughts and high temperatures in the years to come. This could lead to land degradation and agricultural expansion. By 2050, almost 40 percent of the land currently under low-impact agriculture could be converted to more intensive agricultural use, forcing poor farmers to open up ever more marginal lands. One study estimates that climate change could lead to a 50 percent reduction in crop yields for rainfed agricultural crops by 2020. According to crop-climate models, in tropical countries even moderate warming could reduce cereal yields significantly because many crops are already at the limit of their heat tolerance. The areas most vulnerable to climate change—centered in South Asia and Sub-Saharan Africa—also have the largest number of rural poor and rural populations dependent on agriculture. Recent studies show that farming, animal husbandry, informal forestry, and fisheries make up only 7.3 percent of India's gross domestic product (GDP), but these activities constitute 57 percent of GDP of the poor, who are most reliant on natural resources and ecosystem services.

Climate change is likely to aid the spread of invasive alien species, further threatening agricultural productivity and food security through the spread of weeds, pests, and diseases of crops and livestock. The introduction of new and adaptable exotic species to meet growing demands for biofuels, mariculture, aquaculture, and reforestation presents a particular challenge. Ironically, the very characteristics that make a species attractive for introduction under development assistance programs (fast growing, adaptable, high reproductive output, and tolerant of disturbance and a range of environmental conditions) are often the same properties that increase the likelihood of the species becoming invasive. Such events are costly; invasive species accidentally introduced include itch grass, a major weed in cereals in South and Central America, and a range of nematode pests. The economic impacts of invasive alien species can be expensive, costing an estimated $140 billion annually in the United States alone.

Climate change is expected to have serious consequences for water resources. Melting glaciers, higher-intensity and more variable rainfall events, and rising temperatures will contribute to increased inland flooding, water scarcity, and declining water quality. Overall, the greatest human requirement for freshwater resources is for crop irrigation, particularly for farming in arid regions and in the great paddy fields of Asia. In South Asia, hundreds of millions of people depend on perennial rivers such as the Indus, Ganges, and Brahmaputra, all fed by the unique water reservoir formed by the 16,000 Himalayan glaciers. Current trends in glacial melt suggest that the low flows will be substantially reduced as a consequence of climate change, even as the demand for agricultural water is projected to rise by 6 to 10 percent for every 1° C rise in temperature. As a result, even under the most conservative climate projections, the net cereal production in South Asian countries is likely to decline by 4 to 10 percent by the end of this century.

Municipal water accounts for less than a tenth of human use of water, but clean drinking water is a critical need. Today, half of the world's population lives in towns and cities, and one-third of this urban population lacks clean drinking water. These billion have-nots are distributed unevenly across the globe: 700 million city dwellers in Asia, 150 million in Africa, and 120 million in Latin America and the Caribbean. In recent years, governments and city councils have begun to seek opportunities to offset or reduce some of the costs of maintaining urban water supplies—and, perhaps even more important, water quality—through the management of natural resources, particularly forests and wetlands. Most protected areas are established to protect their biodiversity values, but many could be justified on the basis of the other ecosystem services that they provide. From China to Ecuador and Kenya to Mexico, protected areas in forest watersheds safeguard the drinking supplies for some of the world's major cities. In Indonesia, the Gunung Gede Pangrango National Park, for instance, safeguards the drinking water supplies of Jakarta, Bogor, and Sukabumi and generates water with an estimated value of $1.5 billion annually for agriculture and domestic use, while Kerinci-Seblat National Park safeguards water supplies for more than 3.5 million people and 7 million hectares of agricultural land.

Bank projects and programs are already supporting biodiversity conservation and protecting natural habitats and ecosystem services, thereby contributing to effective mitigation and adaptation strategies. Pilot projects that integrate protection of natural habitats and "green" infrastructure into watershed management, flood control, and coastal defense already demonstrate the cost-effectiveness of such ecosystem-based approaches.

Climate change highlights the need to replicate and scale up such interventions, including the following:

- Protecting terrestrial, freshwater, and marine ecosystems and ecological corridors to conserve terrestrial and aquatic biodiversity and ecosystem services
- Integrating protection of natural habitats into strategies to reduce vulnerability to natural disasters (such as floods and cyclones)
- Scaling up country dialogue and sector work on the valuation of ecosystem services and the role of natural ecosystems, biodiversity, and ecosystem services in underpinning economic development
- Emphasizing the linkages between protecting natural habitats and regulating water flows and water quality for agriculture, food security, and domestic and industrial supplies
- Scaling up investments for protected areas and ecosystem services linked to sector lending, such as infrastructure, agriculture, tourism, water supply, fisheries, and forestry
- Promoting greater action on management of invasive alien species, which are linked to land degradation and threaten food security and water supplies
- Emphasizing the multiple benefits of forest conservation and sustainable forest management (carbon sequestration, regulation of water quality, protection from natural hazards, alleviation of poverty, conservation of biodiversity)

- Promoting investments in natural ecosystems as a response to mitigation (avoided deforestation) and adaptation (wetland services)
- Integrating indigenous crops and traditional knowledge on agro-biodiversity and water management into agricultural projects as part of adaptation strategies
- Promoting more sustainable natural resource management strategies linked to agriculture, land use, habitat restoration, forest management, and fisheries
- Developing new financing mechanisms and integrating ecosystem benefits into new adaptation and transformation funds
- Using strategic environmental assessments as tools to promote protection of biodiversity and ecosystem services
- Monitoring investments in ecosystem protection within mainstream lending projects and documenting good practices for dissemination and replication
- Developing new tools to measure the benefits of integrated approaches to climate change (ecosystem services, biodiversity conservation, carbon sequestration, livelihood co-benefits, and resilience).

Promoting further integration of ecosystem-based approaches into responses to climate change and national adaptation strategies will require much greater access to sources of funding, including capitalizing on opportunities to protect natural ecosystems as part of major energy and infrastructure projects. The Bank is facilitating the development of market-based financing mechanisms and piloting new avenues to deepen the reach of the carbon market. New initiatives and investment funds such as the Forest Carbon Partnership Facility, the Forest Investment Program, and the Pilot Program for Climate Resilience provide exciting opportunities to protect natural capital, benefit communities, and use cost-effective green technology to address the challenges of climate change.

CHAPTER 1

The World Bank and Biodiversity Conservation: A Contribution to Action for Climate Change

CLIMATE CHANGE IS A SERIOUS ENVIRONMENTAL CHALLENGE that could undermine the drive for sustainable development. The global mean surface temperature has risen an average of 0.6° C (degree Celsius) over the last 100 years, largely due to the accumulation of greenhouse gases (GHGs) in the atmosphere (Gitay and others 2002). Most of this change has occurred in the last 30 to 40 years, and the rate of increase is accelerating. These rising temperatures will have significant impacts at a global scale and at local and regional levels. Although reducing GHG emissions and reversing climate change are long-term goals, many of the impacts of climate change are already in evidence. As a result, governments, communities, and other sectors of civil society are increasingly concerned with anticipating the future effects of climate change, while searching for strategies to mitigate and adapt to both its current and future effects.

The World Bank's mission is to alleviate poverty and support sustainable development. The conservation and sustainable use of natural habitats and biodiversity will contribute to these goals by protecting ecosystem services that are critical to fulfilling these objectives. Biodiversity is the foundation and mainstay of agriculture, forests, and fisheries. Biological resources provide the raw materials for livelihoods, sustenance, medicines, trade, tourism, and industry. Genetic diversity provides the basis for new breeding programs, improved crops, enhanced agricultural production, and food security. Forests, grasslands, freshwater, and marine and other natural ecosystems provide a range of services often not recognized in national economic accounts but vital to human welfare: regulation of water flows and water quality, flood control, pollination, decontamination, carbon sequestration, soil conservation, and nutrient and hydrological cycling. Sound ecosystem management provides countless benefits to, and opportunities for, human societies, while also supporting the web of life. Ecosystem services and biodiversity conservation contribute to environmental sustainability, a critical Millennium Development Goal and a central pillar of World Bank assistance.

Research undertaken as part of the Millennium Ecosystem Assessment showed that, over the past 50 years, human activities have changed ecosystems more rapidly and extensively than at any comparable period in our history. These changes have contributed to the achievement of many net development gains, but at growing environmental and social costs: habitat loss, land degradation, and reduced access to adequate water and natural resources for many of the world's poorest people. Climate change is likely to compound this environmental degradation.

Terrestrial and oceanic ecosystems play a significant role in the global carbon cycle (see figure 1.1). About 100 gigatons of carbon (GtC) annually are taken up and released by terrestrial ecosystems, and another 100 GtC are taken up and released by marine systems (Matthews and others 2000). These natural fluxes are large compared to the approximately 6.5 GtC emitted annually from fossil fuels and industrial processes and another 1–2 GtC per year emitted as a result of deforestation, predominantly in the tropics (Gitay and others 2002). Natural habitats are a net sink of carbon. Worldwide, soils alone store almost 2,000 GtC (Matthews

FIGURE 1.1
Approximate Stores (Gigatons) and Fluxes (Gigatons per Year) of Carbon

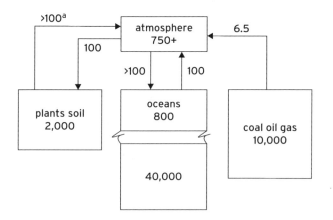

Source: Woods Hole Institute (http://www.whrc.org/carbon/index.htm).
Note: Numbers in the boxes are carbon stores. Numbers beside the arrows are carbon fluxes.
a. Deforestation contributes 1–2 GtC per year.

and others 2000). Furthermore, terrestrial ecosystems remove an estimated 3 GtC, and oceans another 1.7 GtC, of carbon from the atmosphere every year. Appropriate management of terrestrial and aquatic habitats can, therefore, make a significant contribution to reducing GHGs.

Impacts of Climate Change on Ecosystems and Biodiversity

Habitat loss and fragmentation, overexploitation, pollution, the impact of invasive alien species, and, increasingly, climate change all threaten the biological resources and ecosystem services on which humankind depends. Larger concentrations of atmospheric carbon dioxide (CO_2), higher land and ocean temperatures, changes in precipitation, and rise in sea level will affect both natural systems and human welfare. Global warming and climate changes have already had observed impacts on natural ecosystems and species (Moritz and others 2008; van Zonneveld and others 2009). Wetlands, mangroves, coral reefs, cloud forests, and Arctic ecosystems are particularly vulnerable. Climate change is also expected to increase the likelihood of species extinctions and may affect the distribution, behavior, and reproduction of species, patterns and frequency of migrations, as well as intensity of outbreaks of pests and diseases, all of which are likely to affect crop production, food security, and human health.

Some of the most threatened ecosystems globally are Mediterranean-type habitats such as those found in the Cape Floristic Region (CFR), Mediterranean Basin, and southern Chile. The CFR is the smallest of the world's six floral kingdoms, protecting unique Mediterranean-type vegetation known as fynbos. The CFR covers an area of 90,000 square kilometers and is the only floral kingdom to be located entirely within the geographic confines of a single country, South Africa. The CFR contains 9,600 species of vascular plants, many of them endemic; it has been identified as one of the world's "hottest" biodiversity hot spots (see also box 1.1). The rich biodiversity of the CFR is under serious threat as a result of the conversion of natural habitat to permanent agriculture and to rangelands for cattle, sheep, and ostriches; inappropriate fire management; rapid and insensitive infrastructure development; overexploitation of coastal resources and wildflowers; and infestation by alien species. Some important habitats have already been reduced by more than 90 percent, while less than 5 percent of the lowlands enjoys any conservation status. Climate change will exacerbate the threats to these threatened ecosystems and put increasing pressure on water resources, while increasing the vulnerability to fire and the spread of invasive alien species. Maintaining ecological connectivity and reducing further degradation of habitat will be critical strategies for protecting biodiversity and ecosystem services.

Climate change is likely to accelerate the ongoing impoverishment of global biodiversity and degradation of ecosystems caused by unsustainable use of natural

> **BOX 1.1**
> ## Monitoring the Impact of Climate Change in a Biodiversity Hot Spot
>
> The Succulent Karoo biome, which covers 116,000 square kilometers of desert along the Atlantic coast of South Africa and southern Namibia, supports the world's richest succulent flora. It is one of the world's 34 biodiversity hot spots, regions that are the richest in endemic species and also the most threatened on Earth. Together these hot spots harbor more than 75 percent of the most threatened mammals, birds, and amphibians, yet they have already lost more than 85 percent of their original habitat cover. These critical areas for biodiversity are also home to millions of people who are highly dependent on healthy ecosystems for their livelihoods and well-being.
>
> This transboundary area—comprising the Richtersveld, Gariep River, Ais-Ais, and the Fish River Canyon—has a staggering 2,700 plant species, of which 560 are endemic. Compared with vegetation in other hot spots, vegetation in the Richtersveld remains relatively intact in spite of pressures from overgrazing and diamond mining. In recognition of these values, the United Nations Educational, Scientific, and Cultural Organization (UNESCO) has put the Richtersveld cultural and botanical landscape on its World Heritage List.
>
> The area is globally recognized as an example of a biodiversity hot spot under apparent and imminent threat from climate change. Projected time frames for the onset of significant impacts vary from 30 to 50 years, although some botanists believe that early signs of global warming may already be evident in the higher mortalities of *Aloe* species in the Richtersveld. The implications of climate change for ecosystems and livelihoods are highly significant.
>
> Given expected climate change scenarios and the fact that 75 percent of the land is under communal management, a Global Environment Facility (GEF)-funded project in the Richtersveld has opted for a three-tier conservation strategy: (1) forward planning by integrating biodiversity into land use management planning; (2) improved reactive management and implementation of environmental management plans for livestock and mining; and (3) monitoring of the effectiveness of land use planning and management in achieving conservation objectives (for example, monitoring the distribution of *Aloe pillansii* as an indicator species for climate change).
>
> More specifically, the unique attributes of the Richtersveld make the region highly suitable as an international ecological research location for the study of global climate change. The South African research community is developing a network of long-term ecological research sites that act as ecological observatories for change in ecosystems. In this context, the people of the Richtersveld are in the process of forming research partnerships to study global climate change. Specific attention will be given to designing a network of protected areas resilient to species loss. Maintaining ecological connectivity and preventing the degradation of habitat are essential "lines of defense" against the impacts of climate change.

capital and other environmental stresses. Permafrost melt in Mongolia, for instance, is exacerbating the effects of habitat degradation caused by overgrazing and affecting water resources and other ecosystem services (see box 1.2). Similarly, the warming of coastal waters, coral die-off, and impacts on coastal fisheries caused by climate change are worsening the impacts on marine systems of overexploitation by industrial and artisanal fisheries as well as pollution from ships' waste and land sources. Such degradation and disturbance in terrestrial and aquatic ecosystems generate niches that can be exploited by invasive alien species, leading to further ecosystem change and degradation.

> BOX 1.2
> ## Climate Change and Biodiversity Loss in Hövsgöl National Park, Mongolia
> Hövsgöl National Park is centered on Lake Hövsgöl, lying at 1,700 meters above sea level in the mountains of northern Mongolia. Here the winters are long and vicious, with temperatures dropping to below -40° C. The Lake Hövsgöl area lies at the southern edge of the taiga forest and is underlain by permafrost (layers of frozen soil). The region is used by traditional graziers and their livestock. Uncontrolled grazing by sheep, goats, and cattle on the mountain slopes around the lake and the gathering of fuelwood have caused the forest edge to retreat. This loss of forest exposes the ground to sunlight. As a result, the permafrost melts at a faster rate than normal, and aerobic decomposition occurs, producing GHGs. The average temperature in the region rose about 1.4° C during the last 35 years.
>
> In 2001 the Mongolian Academy of Sciences received a five-year GEF grant to study the dynamics of biodiversity loss and permafrost melt in Hövsgöl National Park. The research determined that the active-layer thickness of the permafrost in the Hövsgöl region varies in association with the pressures from livestock grazing. Removal of vegetation cover increases mean summer surface and ground temperatures, accelerating the rate of permafrost melt. The researchers concluded that the impacts of climate change on the steppe and forests are very similar to, and magnify, those caused by nomadic pastoralism and forest cutting. To mitigate these effects, herders need to change grazing strategies to adapt to changing conditions in this harsh and fragile environment. The conclusions regarding land use practices have been summarized in a herders' handbook, which includes recommendations for more rotational grazing to reduce pressure and improve range management. Although little can be done to alter the immediate course of climate change, protecting vegetation cover through appropriate land use practices can slow the rate of permafrost melt and help to protect Mongolia's water resources, biodiversity, and natural ecosystems. These lessons are also relevant to other areas within the temperate mountain-forest-grassland mosaics that stretch from Eastern Europe to eastern Russian Federation and northern China.

Impacts on Human Communities and Livelihoods

Habitat loss and degradation will also increase human vulnerability to climate change. For example, climate change will affect the physical and biological characteristics of coastal areas, modifying the structure and functioning of the ecosystem. As a result, coastal nations face losses of marine resources and fisheries as well as shoreline habitats such as wetlands and mangroves. Rising ocean temperatures cause corals to bleach and, under sustained warm conditions, to die. Nearly 30 percent of warm-water corals in the Caribbean have disappeared since the beginning of the 1980s, a change largely due to increasingly frequent and intense periods of warm sea temperatures. The increase of CO_2 in the atmosphere is also resulting in the acidification of oceans, affecting the calcification of reef plants and animals, especially corals, and thus reducing the ability of reefs to grow vertically and keep pace with rising sea levels. The drowning of atolls and the destruction of corals have long-term implications for coastal zone protection, ecosystem integrity, ecosystem services, and productivity of the tropical seas and fisheries.

Climate change, sea-level rise, and more frequent extreme weather events such as hurricanes will have repercussions on coastal development, water supply, energy, agriculture, and health, among other sectors. The Intergovernmental Panel on Climate Change (IPCC) has assessed the likely regional impacts of climate change (see box 1.3). Table 1.1 shows potential climate-related threats in different Bank

---

BOX 1.3
Likely Regional Impacts on Human Communities and Livelihoods

The fourth assessment of the IPCC studied and reported on the likely regional impacts. The magnitude and timing of impacts will vary with the amount and rate of climate change.

In *Africa*, by 2020, between 75 million and 250 million people are projected to be exposed to increased water stress due to climate change, and, in some countries, yields from rain-fed agriculture could be reduced up to 50 percent. Toward the end of the century, the projected rise in sea level will affect low-lying coastal areas with large populations. The cost of adaptation could amount to at least 5–10 percent of gross domestic product (GDP). By 2080, arid and semiarid land is projected to increase 5–8 percent.

In *Asia*, by the 2050s, the availability of freshwater in Central, South, East, and Southeast Asia, particularly in large river basins, is projected to decrease. Coastal areas, especially heavily populated delta regions in South, East, and Southeast Asia, will be at greatest risk due to increased flooding from the sea and, in some mega-deltas, from the rivers. Climate change is projected to compound the pressures on natural resources and the

client countries, many of them among the world's poorest nations. Many countries will suffer even if the sea level rises only 1 meter, a conservative estimate. A more dramatic rise of up to 5 meters would have even greater impacts, flooding large areas in the Philippines, Brazil, República Bolivariana de Venezuela, Senegal, and Fiji as well as the lower-lying islands and coastal states.

The impacts of climate change in Latin America and the Caribbean have been studied in some detail (Vergara 2005). They include a potential sea-level rise that threatens coastal habitats and human settlements; higher sea surface temperatures; melting of tropical glaciers and snowcaps; warming and drying out of moorlands and other high-altitude ecosystems in the Andes; higher frequency and distribution of forest fires; the spread of tropical disease vectors into the Andes piedmont; changes in agricultural productivity; and impacts on coastal and watershed ecosystems. These changes will have major impacts on the region's rich biodiversity and ecosystem services as well as on human health and livelihoods.

The biophysical implications of sea-level rise will vary greatly in different coastal zones, depending on the nature of coastal landforms and ecosystems. For example, flooding conditions in the pampas in the province of Buenos Aires will be exacerbated by any degree of sea-level rise because of the reduced effectiveness of the natural drainage system. Some coastal areas in Central America and on the Atlantic coast of South America, such as the river deltas of the Magdalena in Colombia, will be subject to inundation risk, as will the large, flat deltas of the Amazon,

---

environment associated with rapid urbanization, industrialization, and economic development. Endemic morbidity and mortality due to diarrheal disease, primarily associated with floods and droughts, are expected to rise in East, South, and Southeast Asia.

In *Latin America*, by mid-century, increases in temperature and associated decreases in soil water are projected to lead to the gradual replacement of tropical forest by savanna in eastern Amazonia. Similarly, areas of semiarid vegetation will tend to be replaced by dryland vegetation. Significant biodiversity will be lost as a result of species extinction in many areas of tropical Latin America. Changes in rainfall patterns and the disappearance of glaciers are projected to reduce the availability of water for human consumption, agriculture, and energy generation.

In *small islands*, sea-level rise is expected to exacerbate inundation, storm surge, erosion, and other coastal hazards. By 2050, climate change is expected to reduce water resources in many small islands, such as in the Caribbean and Pacific, to the point where they become insufficient to meet demand during periods of low rainfall. With higher temperatures, increased invasion by non-native species is expected to occur, particularly on mid- and high-latitude islands.

TABLE 1.1
Five Climate Threats and the Countries Most at Risk

Threat	Low Income	Middle Income	High Income
Drought	Chad, Eritrea, Ethiopia, India, Kenya, Malawi, Mauritania, Mozambique, Niger, Sudan, Zimbabwe	Islamic Republic of Iran	None
Flood	Bangladesh, Benin, Cambodia, India, Lao PDR, Mozambique, Pakistan, Rwanda, Vietnam	China, Sri Lanka, Thailand	None
Storm	Bangladesh, Haiti, Madagascar, Mongolia, Vietnam	China, Fiji, Honduras, Moldova, Philippines, Samoa, Tonga	None
Coastal	Bangladesh, Mauritania, Myanmar, Senegal, Vietnam	China, Arab Republic of Egypt, Indonesia, Libya, Mexico, Tunisia	All low-lying island states
Agriculture	Ethiopia, India, Malawi, Mali, Niger, Pakistan, Senegal, Sudan, Zambia, Zimbabwe	Algeria, Morocco	None

Orinoco, and Paraná rivers. Estuaries such as the Río de la Plata will also suffer increasingly from saltwater intrusion, creating problems in freshwater supply. Potential changes from a rise in sea level reported for the Caribbean Basin range from 3 to 8 millimeters in three years. Such changes will affect both human populations and natural ecosystems. Anticipated increases will threaten aquifer-based freshwater supplies through saline intrusion in many of the smaller islands and lead to flooding of coastal zones. This is a major concern, given that more than 50 percent of the people in most Caribbean states reside within 2 kilometers of the coast. Resources critical to island and coastal populations—including beaches, wetlands, freshwater, fisheries, coral reefs and atolls, and wildlife habitat—are all at risk.

At the other end of the altitudinal spectrum, climate change is affecting mountain ecosystems. Glacial retreat in the Andes is occurring at an alarming rate. Recent measurements show catastrophic declines in the volume of glaciers; these changes are likely to have substantial impacts on water flows to Andean valleys. At lower mountain altitudes, the changes observed include loss of water regulation, increased likelihood of flash fires, and changes in the composition and resilience of ecosystems. Moreover, as temperatures rise, there is a substantive risk of recurring glacial overflows caused by melting ice, placing downstream populations and infrastructure at imminent risk. Warming is also affecting the moorlands, high-altitude ecosystems that are storage areas for water and soil carbon. Climate change will be more pronounced in high-elevation mountain ranges, which are warming faster than adjacent lowlands. Hydrological and ecological changes of this magnitude will result in a loss of unique biodiversity, as well as a loss of many of the ecosystem goods and services provided by these mountains, especially water supply, basin regulation, and associated hydropower potential.

Climate change is expected to affect the supply of, and demand for, water resources as well as environmental flows. All freshwater ecosystems will face ecologically significant impacts by the middle of this century. There will be no "untouched" ecosystems, and the key ecological characteristics of many water bodies are likely to be profoundly transformed, including flow regime, patterns of thermal stratification, and propensity to cycle between oligotrophic (nutrient-poor) and eutrophic (nutrient-rich) states. While aquatic life depends on both the quantity and quality of water, changes in flows are of particular concern because they govern so many ecosystem processes. Many tropical regions, for instance, experience flooding in the wet season and low or no flow during the dry season. In temperate latitudes, spring sees high water following snowpack melt. However, these "normal" patterns can mask the amount of "normal variability" in environmental flows from one year to the next. Eastern Africa, for instance, typically shows interannual variability of 30 percent, so a very wet year can be followed by a very dry one. The Amazon sees little variability between years, but the Pantanal to the south shows relatively large swings. In most regions, climate change is increasing the amount of interannual variability—more droughts or more floods, more very hot days, more intense precipitation—which has a big impact on environmental flows, local agriculture, and human livelihoods (Matthews and others 2009).

Why Protecting Ecosystems and Biodiversity Matters in a Changing World

Current efforts to address climate change focus mainly on reduced emissions of GHGs through cleaner energy strategies and on improved infrastructure to meet new demand for energy and water and to reduce the vulnerability of communities at risk. Both of these approaches are necessary. Nevertheless, in many countries, including the poorest nations, these responses could, and should, be complemented by greater emphasis on natural capital and ecosystem-based approaches to mitigation and adaptation, through improved conservation and more sustainable management of natural habitats and resources.

Improved ecosystem management can enhance resilience to climate change, protect carbon stores, and contribute to adaptation strategies. Climate change is already affecting ecosystems and livelihoods, but enhanced protection and management of biological resources can mitigate these impacts and contribute to solutions as nations and communities strive to adapt. Such ecosystem-based strategies can offer cost-effective, proven, and sustainable solutions to climate change, contributing to, and complementing, other national and regional adaptation strategies.

Protecting forests, wetlands, coastal habitats, and other natural ecosystems can provide social, economic, and environmental benefits, both directly through more sustainable management of biological resources and indirectly through protection of ecosystem services. Protected areas, and the natural habitats within them, can protect watersheds and regulate the flow and quality of water, prevent soil erosion, influence rainfall regimes and local climate, conserve renewable harvestable resources and genetic reservoirs, and protect breeding stocks, natural pollinators, and seed dispersers, which maintain ecosystem health. Floodplain forests and coastal mangroves provide storm protection and serve as safety barriers against natural hazards such as floods, hurricanes, and tsunamis. Natural wetlands filter pollutants and serve as nurseries for local fisheries. Better protection and management of key habitats and natural resources can benefit poor, marginalized, and indigenous communities by maintaining ecosystem services and maintaining access to resources during difficult times, including in times of drought and disaster.

The World Bank Group is a major global funder of biodiversity initiatives, providing support to 598 projects in more than 120 countries during the last 20 years. This portfolio of biodiversity projects represents more than $6 billion in biodiversity investments, including Bank contributions and leveraged co-financing (see table 1.2). Many of those projects are already promoting sound management of natural resources that could contribute to mitigation and adaptation by maintaining and restoring natural ecosystems, improving land and water management, and protecting large blocks of natural habitats across altitudinal gradients. Improved protection of high-biodiversity forests, grasslands, wetlands, and other natural habitats provides benefits for livelihoods as well as carbon storage.

Bank projects directly support biodiversity conservation and sustainable use in a range of natural habitats, from coral reefs to some of the world's highest mountains and from tropical evergreen and monsoon forests to savannas, grasslands, and unique drylands, limestone, marine, and freshwater ecosystems. Many are in centers of recognized global importance for biodiversity: mega-diversity hot spots, remaining wilderness areas, the Global 200 ecoregions described by World Wide Fund for Nature (WWF), and the areas designated as Endemic Bird Areas (EBAs) and as Important Bird Areas (IBAs). Many projects are in countries and regions where communities are most vulnerable to the impacts of climate change. By promoting investments in these locations, the Bank is helping client countries to meet the 2010 targets of the Convention on Biological Diversity (CBD) and to prepare for the impacts of climate change.

A substantial amount of Bank biodiversity funding has been dedicated to protected areas, but there is an increasing focus on improving natural resource management and on mainstreaming biodiversity conservation into forestry, coastal

TABLE 1.2
Total Biodiversity Investments, by Year and Source of Funding
US$ millions

Fiscal Year	World Bank Group						Co-financing	Total Biodiversity Funding
	GEF	IBRD	IDA	Trust Funds	Carbon Finance	Total		
1988	0.00	3.79	2.86	0.00	0.00	6.65	8.95	15.60
1989	0.00	3.16	3.93	0.00	0.00	7.09	5.21	12.30
1990	0.00	129.26	14.22	0.00	0.00	143.48	91.00	234.48
1991	0.00	97.17	35.48	0.00	0.00	132.65	129.94	262.59
1992	23.20	91.21	125.97	0.00	0.00	240.37	130.17	370.55
1993	29.79	17.13	28.37	0.00	0.00	75.29	43.68	118.97
1994	51.27	27.94	54.01	0.00	0.00	133.21	63.95	197.17
1995	44.06	55.81	34.80	36.66	0.00	171.33	176.06	347.40
1996	74.23	40.89	5.07	0.30	0.00	120.48	70.48	190.96
1997	95.90	39.29	103.78	2.00	0.00	240.97	158.46	399.43
1998	78.27	59.64	122.86	0.20	0.00	260.96	252.68	513.64
1999	45.12	15.87	40.15	3.23	0.00	104.36	101.97	206.34
2000	52.07	49.59	14.05	7.35	0.00	123.05	60.74	183.80
2001	166.75	49.54	29.41	27.90	0.00	273.59	268.68	542.27
2002	164.92	15.10	55.49	5.67	0.00	241.18	205.21	446.39
2003	81.31	33.33	62.29	0.00	0.00	176.92	110.68	287.60
2004	103.46	38.95	66.60	4.42	0.44	213.87	274.97	488.84
2005	118.63	88.64	73.20	14.46	0.00	294.93	154.38	449.31
2006	156.02	78.65	25.39	17.70	19.20	296.96	172.33	469.29
2007	70.61	35.54	27.52	3.02	1.04	137.73	55.78	193.51
2008	48.36	33.38	0.80	1.10	0.00	83.64	178.11	261.75
Total	1,403.95	1,003.86	926.23	124.00	20.68	3,478.72	2,713.45	6,192.18

Source: World Bank 2008a.
Note: GEF = Global Environment Facility; IBRD = International Bank for Reconstruction and Development; IDA = International Development Association.

zone management, and agriculture. Beyond these "traditional" natural resource sectors, the Bank has successfully tested modalities for supporting protection and improved management of natural habitats through Bank-funded energy and infrastructure projects and development policy lending. The Bank is also developing innovative climate investment funds, including funds that will target natural ecosystems, especially forests, as carbon stores.

The global focus on climate change, and the need to address likely impacts at the national level, provides a new imperative to protect the natural capital and ecosystem services on which many communities depend. The Bank's access to lending resources and multiple financing instruments provides opportunities to promote ecosystem-based approaches to climate change within national agendas as a critical part of sustainable development. Such efforts would complement assistance to clients in developing adaptation strategies as well as ongoing dialogues on governance and improved natural resource management. The new multidonor climate investment funds, described in chapter 5, provide exciting opportunities to protect habitats and ecosystem services, while addressing the climate change agenda.

This book offers a compelling argument for including ecosystem-based approaches as an essential third pillar in national strategies to address climate change. Many of the case studies presented in boxes derive from lessons learned and best practice in Bank projects. Natural ecosystems can contribute to strategies to reduce GHG emissions and can complement infrastructure investments to reduce vulnerability to climate change. Chapter 2 examines the role of natural ecosystems as carbon stores and sinks. It also provides information and examples of how effective conservation can contribute to low-technology, low-cost mitigation. Chapter 3 demonstrates how integrating protection of natural habitats and management of natural resources into adaptation plans can contribute to cost-effective strategies for reducing vulnerability to climate change. Chapter 4 emphasizes the links between ecosystem services and human livelihoods, agriculture, and water. The final chapter provides an overview of available financing instruments to support ecosystem-based approaches to climate change, including climate investment funds and the larger carbon market.

CHAPTER 2

Natural Ecosystems and Mitigation

CLIMATE CHANGE IS ALREADY AFFECTING NATURAL SYSTEMS, weather events, and lives and livelihoods. The current level of greenhouse gases (GHGs) in the atmosphere is equivalent to approximately 430 parts per million of carbon dioxide (CO_2e), which is almost double the amount before the Industrial Revolution (Stern 2007). If emissions remain at current rates, by 2050 the concentrations of GHGs in the atmosphere will reach 550 parts per million and continue to increase thereafter. While emissions from fuel are the main culprits, changes in land use also make a significant contribution to overall levels of GHGs.

It is highly likely that the global average rise in temperature associated with this GHG concentration would be above 2° C. As shown in figure 2.1, such changes in temperature would have adverse effects on food security, water availability, weather conditions, and species diversity and have severe effects on ecosystems such as coral reefs. Therefore, it is extremely important for countries to mitigate climate change and reduce GHG emissions to a level that Earth's natural sinks can balance. According to the fourth assessment report of the Intergovernmental Panel on Climate Change (IPCC 2007), low to medium stabilization levels (450–550 parts per million of CO_2e) would prevent drastic harm to ecosystems and human livelihoods but would be achievable only through concerted global efforts. Immediate implementation of mitigation measures is, therefore, essential to meet these emission goals. Biodiversity and natural ecosystems, with their

FIGURE 2.1
Likely Changes to Earth Systems Depending on Mitigation Activities Undertaken

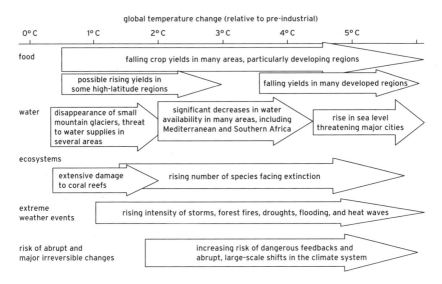

Source: Stern 2007.

vast capacity to store carbon and regulate the carbon cycle, can play a key role in such mitigation efforts.

Mitigation involves reducing GHG emissions from energy-related or land use changes and enhancing natural GHG sinks. Biological mitigation of GHGs can occur through (a) sequestration by increasing the size of carbon pools (for example, through afforestation, reforestation, and restoration of other natural habitats), (b) maintenance of existing carbon stores (for example, through avoidance of deforestation or protection of wetlands), (c) maintenance of healthy coral reefs and the ocean carbon sink, and (d) substitution of fossil fuel energy with cleaner technologies that rely on biomass. The global potential of biological mitigation options through afforestation, reforestation, avoided deforestation, improved agriculture, and management of grazing land and forests is estimated at 100 gigatons of carbon (GtC) by 2050, which is equivalent to about 10–20 percent of projected fossil fuel emissions during that period.

Afforestation and Reforestation

Under current guidance from the United Nations Framework Convention on Climate Change (UNFCCC) and the Kyoto Protocol's Clean Development Mechanism (CDM), protection of standing forests and other natural habitats is not eligible for carbon credits. Instead, most habitat-related mitigation activities focus on increased sequestration of carbon through afforestation and reforestation projects.

The Bank is involved in afforestation and reforestation efforts throughout the world (see box 2.1). These projects promote carbon sequestration but are often linked to maintenance of other ecosystem services and local benefits, such as watershed protection or provision of fuelwood and fodder. Similarly, the World Bank, through the BioCarbon Fund, is financing reforestation of more than 23,000 hectares of *Acacia senegalensis*, a species native to the African Sahel, on degraded communal

---

BOX 2.1
Reforestation under the BioCarbon Fund

Brazil: Reforestation around hydro reservoirs

Natural forests will be restored on approximately 5,576 hectares of land around four reservoirs created by hydroelectric plants in the state of São Paulo. Planting a mix of at least 80 native species will regenerate forested areas, protect the recreational use of the area, and improve the value of the lands for tourism. Many of the targeted sites are connected to existing forested areas and linked to riverine habitats. Restoration of forest is expected to sequester 0.67 Mt CO_2e by 2012 and 1.66 Mt CO_2e by 2017, increasing critical habitats, creating vital wildlife corridors, and connecting the newly forested lands with existing conservation areas.

China: Pearl River watershed management

This project is reforesting 4,000 hectares in the Guangxi Zhuang Autonomous Region, which includes half of the Pearl River basin and is an area of high biodiversity value. The sites selected for planting are shrubland, grassland, and areas with less than 30 percent tree cover; 75 percent of the species planted will be native. *Eucalyptus*, grown in China for a century, will make up most of the exotics. The restoration of forests along the middle and upper reaches of the Pearl River will serve as a demonstration model for watershed management. Carbon sequestered by a plantation will be used as a cash crop and will generate income for local communities. As the first life-size Land Use, Land Use Change, and Forestry (LULUCF) project in China, it will also test how afforestation activities can generate reductions in GHG emissions that can be measured, monitored, and certified. The reforested land is expected to sequester around 0.34 Mt CO_2e by 2012 and around 0.46 Mt CO_2e by 2017, along with restoring forest connectivity between two nature reserves (Mulun and Jiuwandashan in Huanjiang County) to provide a wildlife corridor for animal movements.

(continued)

> **BOX 2.1** *(continued)*
> **Kenya: Green belt movement**
>
> This project is reforesting 4,000 hectares of degraded public and private lands with high community access in the Aberdare range and Mount Kenya watersheds. These forests host a large number of threatened fauna species and are internationally recognized as an Important Bird Area. Although many of these forests are officially protected as a reserve, they are threatened by illegal logging and cultivation. The project will pay local communities and provide them with the technology and knowledge to reforest and manage these lands. Replanting on denuded steep slopes will reduce erosion, protect water sources, and regulate water flows. Communities will be organized into community forest associations that will develop management plans. The long-term goal is to use the regrown forest in a sustainable manner for a variety of products, including fuelwood, charcoal, timber, and medicine, among other uses. Planting of trees on lands around the reserve forests is expected to reduce pressure on remaining natural forests, while the planting of native species will enrich local biodiversity and protect ecosystem services. The project is expected to sequester around 0.1 Mt CO_2e by 2012 and 0.38 Mt CO_2e by 2017.

land throughout Mali and Niger. Plantations of this robust native species will restore habitat for native insects, animals, and birds and is projected to sequester approximately 0.3 megaton of CO_2e (Mt CO_2e) by 2017 and 0.8 Mt CO_2e by 2035 in Mali and 0.24 Mt CO_2e by 2012 and around 0.82 Mt CO_2e by 2017 in Niger. The project will greatly aid the local communities by creating jobs and increasing incomes through the sale of high-quality Arabic gum (see box 2.2).

Afforestation and reforestation projects will affect biodiversity and ecosystem services depending on the land use and ecosystem being replaced and management applied. The reforestation of degraded lands has the potential to produce the greatest benefits for biodiversity, especially with careful selection of species and sites, planting of native species, and efforts to accommodate the needs of native wildlife. Plantations or natural reforestation may contribute to the dispersal capabilities of wildlife by extending areas of forest habitat or providing connectivity between patches of habitat in a formerly fragmented landscape. Even single-species plantations may provide some biodiversity benefits if they incorporate features such as retaining borders of native forest along riverbanks or protecting natural wetlands. In contrast, planting fast-growing exotic species, or species with known potential to become invasive, is likely to provide few biodiversity gains but may provide short-term benefits by reducing soil erosion or providing a ready source of fuelwood and timber.

> **BOX 2.2**
> **Building Resilience by Promoting Native Vegetation in Mali**
>
> For the past 30 years, the Nara area of northern Mali has suffered decreases in rainfall and water levels, land degradation, loss of forest canopy, and change in plant species composition. Tree cutting for firewood, charcoal, and shifting agriculture has been a leading cause of deforestation in the area. The loss of natural vegetation has reduced the resilience of arid zone ecosystems to recurrent droughts. As a consequence of land degradation, the Nara people are facing famine, poverty, and migration. In an already drought-afflicted region, additional climatic stresses are going to be detrimental to food security and development.
>
> Improved management of natural resources and indigenous vegetation can help to build resilience against climate change and contribute to more sustainable livestock husbandry and farming. The BioCarbon Fund is providing funding for reforesting 6,000 hectares of *Acacia senegalensis*, a species endemic to the African Sahel. It is superbly adapted to harsh ecological conditions and produces several environmental benefits. Besides producing gum, it enables the rehabilitation of degraded areas that have become unfit for agriculture. *Acacia's* powerful root system makes it efficient for fixing dunes and controlling wind and water erosion. Its nitrogen-fixing ability improves soil fertility. Local organizations and farmers in the Nara region will develop and manage cost-effective modern nurseries, plant trees, maintain plantations, and harvest Arabic gum. The project will also diversify agricultural activities through intercropping with groundnuts and cowpeas. By restoring healthy populations of *Acacia senegalensis*, the project will also benefit local biodiversity and provide more fodder for local cattle.
>
> The project is expected to sequester around 0.3 Mt CO_2e by 2017 and 0.8 Mt CO_2e by 2035. Overall the project is expected to create about 1,700 jobs for plantation management and the production, transport, and selling of Arabic gum. The management of the nurseries will create another 200 jobs. Some 10,000 farming families are expected to benefit from the project with their own *Acacia* plantations (approximately 1 hectare per participant). Hundreds of farming families are expected to benefit from the additional revenues generated by Arabic gum, grains, and forage, combined with payments for reducing emissions (known as certified emissions reductions, or CERs).

Plantations of native tree species will support more biodiversity than exotic species. Plantations of mixed tree species will usually support more biodiversity than monocultures, especially if designed to allow for the colonization and establishment of diverse understory plant communities. Since loss of soil carbon occurs for several years following harvesting and replanting—due to the exposure of soil, increased leaching and runoff, and reduced inputs from litter—long-rotation plantations in which vegetation and soil carbon are allowed to accumulate are more beneficial than short-rotation plantations. Short-rotation forests, with their simpler structure, foster less species richness than longer-lived forests, but products from short-rotation plantations may alleviate harvesting pressure on primary forests.

Securing Carbon Stores through Protection and Restoration of Natural Ecosystems

Many areas of remaining terrestrial habitats and high-biodiversity value overlap areas with large carbon reservoirs. In such biologically important areas, establishment of protected areas or strengthened management can be expected to contribute to the protection of existing carbon reservoirs.

Forests

Forests cover about 30 percent of the world's land area, but they store about 50 percent of Earth's terrestrial carbon (1,150 GtC) in plant biomass, litter, debris, or soil (Watson and others 2000). The relative size of these carbon pools depends on the type of forests and the ecoregions in which they occur (see table 2.1). Land use changes including expansion of human settlements, conversion to agricultural land, and unsustainable logging practices are major threats to forests, resulting in both habitat loss and fragmentation. At the current rate of deforestation, about

TABLE 2.1
Carbon Stocks in Natural Ecosystems and Croplands

Biome	Area[a]	Carbon Stocks		
		Vegetation[b]	Soil[b]	Total[b]
Tropical forests	17.6	212	216	428
Temperate forests	10.4	59	100	159
Boreal forests	13.7	88	471	559
Tropical savannas	22.5	66	264	330
Temperate grasslands	12.5	9	295	304
Deserts and semideserts	45.5	8	191	199
Tundra	9.5	6	121	127
Wetlands	3.5	15	225	240
Croplands	16.0	3	128	131
Total	151.2	466	2,011	2,477

Source: Watson and others 2000.
a. 10 million square kilometers.
b. Gigatons of carbon.

13 million hectares a year (FAO 2005), the world's forests are severely threatened. As these forests are lost, so too are the ecosystem services they provide, including their role as carbon stores and sinks.

About 20 percent of the world's GHG emissions are caused by deforestation and land use changes. The problem is especially acute in the tropics, which include some of the world's most biologically rich countries. In tropical regions, emissions attributable to deforestation and other land clearance are much higher, up to 40 percent of national totals. Brazil and Indonesia together currently account for approximately 54 percent of all emissions from forest loss (Baumert, Herzog, and Pershing 2006). As shown in figure 2.2, some forests with high potential for cash crops also have significant carbon reserves, making these forests and carbon

FIGURE 2.2
Forest Area and Forest Carbon Stocks on Lands Suitable for Major Drivers of Tropical Deforestation

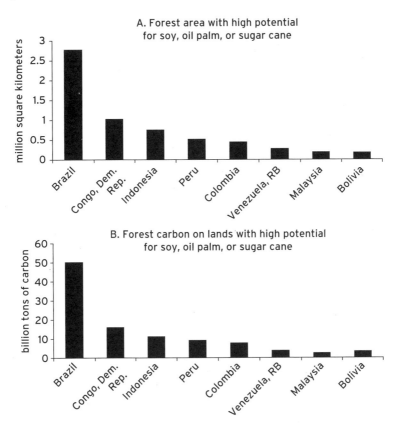

Source: Stickler and others 2007.

reserves highly vulnerable to deforestation activities. Thus most of Indonesia's GHG emissions come from deforestation and land clearance, including clearing and burning of peat swamp forests for agriculture and oil palm production. If current rates of deforestation in Indonesia remain the same through 2012, it is estimated that emissions from this deforestation will equal almost 40 percent of the annual emission reduction targets set for Annex 1 countries under the Kyoto Protocol (Santili and others 2005). Reducing deforestation and forest degradation in key biodiversity countries clearly affords exciting opportunities to address climate change, conservation, and development (see box 2.3).

Key strategies for conserving forests include establishing and strengthening management of protected forests and adopting more sustainable forest management. For example, improvements in sustainable forest management could store an extra 170 Mt CO_2e per year by 2010, or about 3 percent of global CO_2 emissions (Watson and others 2000). Many Bank projects with a focus on improved forest management and protected areas are already contributing to maintaining carbon stores in these forests (see box 2.4). The Russian Federation, for example, contains about 22 percent of the world's forests, including 25 percent of all old-growth forests. These 770 million hectares of forests make up the largest share of temperate and boreal forests among Bank client countries; they harbor important endangered and endemic biodiversity and serve to protect permafrost areas, which are important carbon stores. Because of Russia's large size and extensive forest cover, there is a compelling need to balance economic development in the forest sector with sustainable forest management. Efforts to improve forest and fire management in eastern Russia are retaining important carbon stores in the boreal forests and underlying peatlands, while also protecting the region's rich biodiversity, including tigers, in Khabarovsk Kray.

Efforts to reduce deforestation and degradation have a large role in maintaining carbon stocks in standing forests over the short term. The Thirteenth Conference of the Parties to the UNFCCC in Bali in December 2007 called for rewarding nations and communities for improved protection and management of forests. The ongoing discussion regarding the inclusion of existing forests in international climate mitigation frameworks represents a significant opportunity for both climate and conservation efforts. Acceptance of reducing emissions from deforestation and forest degradation (REDD) as a viable international mechanism for emissions abatement could offer a new platform and financing mechanism for protecting biodiversity, ecosystem services, and forest livelihoods.

The Bank is developing and testing new financing mechanisms to pilot modalities for REDD. The Ankeniheny-Mantadia-Zahamena Corridor Restoration and Conservation Carbon Project is an innovative initiative to conserve and restore the threatened humid forests of Madagascar. The project is promoting natural regeneration and ecological restoration of around 3,020 hectares on degraded

BOX 2.3
Economic Arguments for Sustainable Forest Management

A study from Mount Cameroon comparing low-impact logging with more extreme uses of land found that private benefits favor the conversion of forests to small-scale agriculture. Conversion to oil palm and rubber plantations, however, yielded negative private benefits once the effect of market distortions was removed. Social benefits from nontimber forest products, sedimentation control, and flood prevention were highest under sustainable forestry, as were global benefits from carbon storage. This was true for a range of option, bequest, and existence values. Overall, the total economic value of sustainable forestry was 18 percent greater than that of small-scale farming ($2,570 compared with $2,110 per hectare).

Net Present Value of Various Uses of Tropical Forest in Cameroon

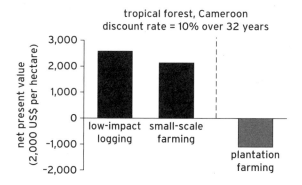

Source: Balmford and others 2002.

land along the buffer zones of two national parks: the Analamazaotra Special Reserve and the Mantadia National Park. By creating sustainable use, the project aims to protect a total area of 425,000 hectares, reducing GHG emissions from deforestation and forest degradation. The reforestation component of the project is expected to sequester around 0.12 Mt CO_2e by 2012 and around 0.35 Mt CO_2e by 2017 (Kyoto compliant), while the avoided deforestation component could generate as much as 4 Mt CO_2e by 2017 (non-Kyoto compliant). Funds from the sale of carbon credits on the voluntary market are being used to finance sustainable livelihood activities in the region, including the planting of fruit tree orchards and fuelwood plantations that will increase farmers' income and reduce pressure on native forests.

> **BOX 2.4**
> **Carbon and Conservation in the Forests of Indonesia**
> In December 2004 a tsunami struck Aceh, Indonesia, causing a large-scale humanitarian crisis, especially along the west coast. In this narrow coastal belt, communities and agricultural lands border directly on protected forests and the karst mountain ranges of the Gunung Leuser National Park and ecosystem in the south and the Ulu Masen Forest Complex in the north. More than two-thirds of the province remains under forests. Even within Indonesia, a mega-diversity country, this area is unique, comprising the largest remaining contiguous forested area (3.3 million hectares) with the richest assemblage of wildlife in Southeast Asia, including tigers, elephants, rhinos, and orangutans. These areas also provide valuable ecosystem services needed for Aceh's recovery, including water supply, flood prevention, erosion mitigation, and climate regulation.
>
> The reconstruction effort raised concerns about how the enormous amount of timber needed for rebuilding could be obtained without endangering these forests. In August 2005, a long-awaited peace accord between the Indonesian government and the Free Aceh Movement effectively removed the barrier to widespread logging activities. Two environmental NGOs, Leuser International Foundation and Flora and Fauna International, both with a long history of working in Aceh, prepared a proposal to the Multi-Donor Fund for the Aceh Forest and Environment Project (AFEP) to ensure the protection of Aceh's forests.

Wetlands

Natural ecosystems are not all equal in their values for biodiversity conservation or their roles in storing carbon and providing other ecosystem services. Various types of wetlands—including swamp forests, mangroves, peatlands, mires, and marshes—are important carbon sinks and stores. Depending on hydrology and vegetation type, both above- and below-ground storage of carbon can be significant in these ecosystems. Anaerobic conditions in inundated wetland soils that slow decomposition rates contribute to long-term soil carbon storage and formation of carbon-rich peats. Such slow decomposition over thousands of years forms peatlands that can extend up to 20 meters in depth and represent some 25 percent of the world pool of soil carbon, an estimated 550 GtC (Parish and others 2008). Peatlands act as carbon sinks, sequestering an estimated 0.3 ton of carbon (tC) per hectare per year, even after accounting for methane emissions (Pena 2009). Moreover, all peatlands, including those in the

> The main objectives of AFEP are to (a) protect the environmental services provided by Aceh's coastal and terrestrial forest ecosystems during, and beyond, the reconstruction and to (b) mainstream environmental concerns in the reconstruction process. AFEP produces accurate and timely information on the state of the province's forests and is building the capacity of the provincial forest and conservation administration. It is helping to develop a model for community-based sustainable forest management and fostering integration of forest and conservation issues into the land use planning process through development of provincial, district, and subdistrict spatial plans. Forest monitoring is carried out at three mutually supportive levels: through remote sensing, aerial surveys, and ground-level community monitoring teams. Aceh's governor, Irwandi, declared a logging moratorium so that new policies and programs could be formulated and implemented. The project's flexible approach to post-disaster, post-conflict reconstruction has benefited from local participation, including collaboration with religious leaders to include environmental and conservation messages in mosque sermons.
>
> The project is also assisting the government of Aceh in developing and promoting REDD assistance for Aceh. A REDD pilot project plan for Ulu Masen achieved the climate, community, and biodiversity standards certification in February 2008. The project is expected to prevent 100 million tons of GHG emissions over the next 30 years by reducing deforestation in Ulu Masen by a staggering 85 percent. The expected 3.3 million carbon credits generated annually will help to finance forest conservation as well as development projects for local villagers, who are some of Indonesia's poorest communities.

boreal zones and Arctic, are refugia for some of the world's rarest species of wetland-dependent flora and fauna.

In recent decades, drainage and conversion for agriculture have led to massive loss of wetland habitats and changed peatlands from a global carbon sink to an emerging source of carbon. Changes in hydrology and reduction in soil saturation expose the soil to air, causing the peat to collapse and the soil carbon to oxidize to carbon dioxide. An estimated 3 billion tC annually, about 10 percent of all reported emissions, are produced as a result of this degradation (Parish and others 2008). Two-thirds of these emissions are concentrated in Southeast Asia, where clearance of swamp forests to expand oil palm plantations and agriculture threaten these unique habitats. Ironically, swamp forests are being cleared in Indonesia to expand the production of oil palm for biofuels.

Working against this trend, Wetlands International has been collaborating with the provincial government and the Indonesian Department of Conservation to establish a new national park in the province of South Sumatra. The Sembilang National Park and adjacent Berbak National Park, Indonesia's first Ramsar site, together protect some of Sumatra's most important remaining lowland forests, including large tracts of peat swamp forests and the most important mangroves in western Indonesia. These areas are important carbon sinks but also provide protection for large mammals (tiger, Sumatran rhino, and tapir), migratory birds, and breeding populations of rare storks. The extensive coastal mangrove swamps also provide critical spawning and nursery grounds for inshore fisheries, an important source of local livelihoods. Thus, conservation efforts, supported through a Global Environment Facility (GEF) project, are contributing to biodiversity and social benefits as well as protecting a major carbon store.

Coastal wetlands, including mangroves, serve as carbon stores and sinks (see box 2.5). Mangroves store as much as 45 tC per hectare (Bouillon and others 2008) and sequester another 1.5 tC per hectare per year (Ong 1993). This amount of carbon sequestration is comparable with that of other tropical forests and may be underestimated due to the lack of information about fine root activities. In addition to carbon sequestration, coastal wetlands provide a wide range of other ecosystem services, including coastal defense, protection against extreme weather events, trapping of sediment, and provision of nutrients and nurseries for coastal fisheries. A study on the Mesoamerican reef, for example, showed that there are as many as 25 times more fish on reefs close to mangrove areas than in areas where mangroves have been cut down. High population pressure in coastal

---

BOX 2.5
Nariva Wetland Restoration and Carbon Offsets in Trinidad and Tobago

The Nariva protected area (7,000 hectares) is one of the most important protected areas in Trinidad and Tobago and is also a Ramsar site. Its varied mosaic of vegetation communities includes tropical rain forest, palm forests, mangroves, and grass savannas. However, these ecosystems have been threatened by hydrological changes arising from a newly constructed water reservoir upstream and more than 10 years (1985–96) of illegal forest clearing by rice farmers.

A Bank project to restore the Nariva wetlands provides a unique opportunity to combine the goals of GHG mitigation with adaptation needs. The project will support carbon sequestration through the reforestation and restoration of the natural drainage regime of the Nariva wetlands ecosystem. Restoration of the wetlands will strengthen their ability to provide a natural buffer for inland areas, representing an adaptation measure to anticipated increases in weather variability.

areas has, however, led to the conversion of many mangrove areas to other uses, including infrastructure, aquaculture, and rice and salt production. Almost 225,000 metric tons of carbon sequestration potential is lost each year because of the current rates of mangrove destruction. In addition to their lost value as a carbon sink, disturbed mangrove soils release more than 11 million metric tons of carbon annually.

Grasslands

Grasslands, including savannas, occur on every continent except Antarctica and constitute about 34 percent of the global stock of terrestrial carbon, most of which is stored in their soil systems. Changes in grassland vegetation due to overgrazing, conversion to cropland, desertification, fire, fragmentation, and introduction of non-native species affect their carbon storage capacity and may, in some cases, even lead to them becoming a net source of CO_2. For example, grasslands may lose 20 to 50 percent of their soil organic carbon content through cultivation, erosion, and degradation. Moreover, burning of biomass, especially in tropical savannas, contributes more than 40 percent of gross global carbon dioxide emissions (Matthews and others 2000).

This loss of carbon storage capacity in grasslands is accompanied by the loss of grassland-dependent birds and herbivore species, leading to biodiversity loss. Approximately 23 of 217 areas designated as Endemic Bird Areas name grassland as the key habitat. In the United States, population trend data over a nearly 30-year period show a constant decline in the number of grassland-dependent

---

Afforestation and reforestation activities on 1,200 hectares of the wetlands is expected to generate carbon credits for approximately 193,000 tons of CO_2e up to 2017, which will be purchased by the BioCarbon Fund. This investment will fund the restoration work, including the following activities:

- Restoration of natural hydrology will help to restore Nariva's ecological functions, including active management of the landscape to ensure the survival of the existing forest as well as reforested areas.
- Between 1,000 and 1,500 hectares are being reforested with native terrestrial and aquatic species. Mechanical and chemical treatment of invasive species may be required to open areas for more natural plant communities.
- A fire management program will protect the newly restored vegetation.
- A monitoring plan will record the impact of reforestation activities and monitor biodiversity, including key species.

species. Similarly, studies show that the population of African herbivores within the protected Serengeti ecosystem has stabilized but the density of herbivores in areas outside the protected area has declined as land conversion has led to loss and degradation of grassland habitats.

Improved management of production in grasslands (for example, management of grazing, creation of protected grasslands and set-aside areas, improved productivity, and fire management) can enhance carbon storage in soils and vegetation, while enhancing other ecosystem services (see box 2.6). Silvopastoral projects in Central America have also demonstrated the economic and ecological benefits of increasing tree cover in cattle pastures. Such agroforestry systems have the potential to sequester carbon, improve livelihoods, and provide functional

---

BOX 2.6
Safeguarding Grasslands to Capture Carbon: Lessons from China

The vast area and wide distribution of China's grasslands suggests that they could have widespread effects on regional climate and global carbon cycles. The Gansu and Xinjiang Pastoral Development Project seeks to produce global environmental benefits by restoring biodiversity and increasing the productivity of grassland resources in the globally significant ecoregions of Tien Shan, Altai Shan, and Qilian Shan. These benefits will result from implementation of participatory grassland management plans delaying and shortening the spring and summer grazing periods in the high mountain grasslands. Reduced grazing pressures will lead to increased species diversity, increased biomass productivity, and improved grazing conditions for wild ungulates as well as herds of sheep and other livestock managed by local herders.

Reduced grazing pressure will provide significant carbon benefits. Improved pasture management practices increase the amount of carbon entering the soil as plant residues, suppress the rate of decomposition of soil carbon, and reduce soil loss due to overgrazing. The project is also promoting more intensive management of lowland pastures, with inputs of inorganic and organic fertilizers, as well as production of livestock foodstuffs to reduce pressure on mountain pastures. Improved practices, such as rotational grazing, include community-based regulation of grazing intensity and frequency. The economic benefits of carbon sequestration were estimated using the shadow price of CO_2 at $20 per tC per year (discounted at a 12 percent interest rate over the 20-year period), which is equivalent to $5.50 per tC. It was estimated that the adoption of better management practices on the pastures would elicit a carbon gain of about 3 to 15 tC per year, depending on the degree of degradation. After three years, carbon benefits from reduced grazing and improved management are expected to increase up to 50 tC per hectare.

links between forest fragments and other critical habitats as part of a broad landscape management strategy for biodiversity conservation.

Protected Areas: A Convenient Solution to Protect Carbon Sinks and Ecosystem Services

Protected areas are the cornerstones of biodiversity conservation and a valuable buffer against the impacts of climate change. They are also a promising tool for reducing emissions from habitat degradation and deforestation, as they generally have well-defined boundaries and incorporate legal restrictions on land use changes. Many protected areas overlay areas of high carbon stocks. Globally ecosystems within protected areas store more than 312 GtC or 15 percent of the terrestrial carbon stock (Campbell and others 2008), but the degree to which carbon stocks are protected varies among regions, as shown in figure 2.3.

Designation of protected areas alone does not guarantee protection of the natural ecosystems within their boundaries. Although many studies show that deforestation is often less within protected areas than in nearby unprotected lands, many reserves have weak or no effective management and suffer from encroachment. Between 2000 and 2005, more than 1.7 million hectares were

FIGURE 2.3
Amount of Carbon Stored in Protected Areas, by Region

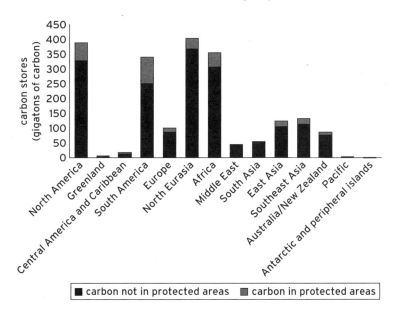

Source: Campbell and others 2008.

cleared within protected areas in the humid tropics alone, that is, 0.81 percent of forest cover was lost (Kapos and others 2008). Globally, more strictly protected areas (International Union for Conservation of Nature [IUCN] management categories I and II) in humid tropical forests showed less forest loss (0.53 percent) than the protected area network overall. Based on these estimates of deforestation, it is estimated that forest loss in protected areas contributed as much as 990 Mt CO_2e emissions between 2000 and 2005 or around 3 percent of total emissions from tropical deforestation. The real level of emissions depends on the use to which the deforested areas are put, for example, arable crops, pasture, and oil palm. The estimated total carbon loss from deforestation within protected areas during 2000–05 was especially high in the Neotropics because of the high carbon content and high rate of deforestation in the Brazilian Amazon.

Protected areas clearly can play an important role in maintaining carbon stores as well as biodiversity, especially if they are well protected and effectively managed. A major share of Bank and GEF biodiversity funding has gone to create sustainable protected area networks, including establishment of new parks and support to strengthen existing protected areas, including promotion of innovative models of management and new financing. Projects include conservation planning and establishment of new protected areas and biological corridors (for example, in Brazil, Central America, Georgia, and Ghana); improved management of "paper parks" and existing protected areas (Bolivia, Ecuador, India, Madagascar, Pakistan, Russia, Uganda); control of invasive exotic plants (Mauritius, the Seychelles, and South Africa); protection and restoration of wetlands and other native habitats (Bulgaria, Croatia, and Indonesia); community management of terrestrial and marine protected areas, indigenous reserves, sacred groves, and clan conservation areas (Colombia, Ecuador, Ghana, Indonesia, Papua New Guinea, Peru, and Samoa); and sustainable finance for protected areas and conservation (Bhutan, Madagascar, Peru, and Tanzania). Large areas of natural habitat are being conserved through transboundary projects in regions such as Central Asia and Mesoamerica, as well as through planning and establishing new protected areas within a mosaic of other improved management systems in the extensive forest wilderness areas of Brazil and Russia.

The Bank's role in supporting biodiversity conservation and protected areas in biologically rich countries could be further optimized by targeting additional carbon funds to areas that have both high biodiversity and high carbon stocks (see box 2.7). In Vietnam, for instance, 58 percent of the areas with high biodiversity overlap with the areas of high carbon stocks, but protected areas cover only about 30 percent of these high-biodiversity lands. Similarly, in Papua New Guinea, only 17 percent of the areas that are high in biodiversity and carbon are within protected areas (Kapos and others 2008). New conservation strategies focusing on ecosystem services as well as biodiversity could focus additional attention and resources on areas where protection would lead to both biodiversity and carbon sequestration benefits.

BOX 2.7
Amazon Region Protected Areas Program: A Storehouse for Carbon and Biodiversity

The Amazon Region Protected Areas (ARPA) Program is an initiative of the Brazilian government to support biodiversity conservation in the Brazilian Amazon, one of the world's largest remaining wilderness areas and an important carbon store. Under the ARPA Program, Brazil has created 22.28 million hectares of protected areas since 2000, surpassing its first-phase target of 18 million hectares. With government support and additional grant funding from the GEF, Kreditanstalt für Wiederaufbau (KfW), and World Wide Fund for Nature (WWF), ARPA has strengthened the management of an additional 8.65 million hectares of existing protected areas. With these 30.93 million hectares of biodiversity-rich forests—a mosaic of state, provincial, private, and indigenous reserves—ARPA is the world's largest protected area program. Plans for the future are even more ambitious: to create a system of well-managed parks and other protected areas, including extractive and indigenous reserves, that encompasses some 50 million hectares, an area larger than the entire U.S. system of national parks.

ARPA was established to protect the rich biodiversity of the Amazon basin, but the mosaic of protected areas contributes to both Brazilian and global efforts to fight climate change through avoided deforestation. The carbon stock in ARPA reserves is estimated at 4.5 billion tC, with potential reductions in emissions estimated at 1.8 billion tC. This role is recognized in the *Stern Review on the Economics of Climate Change* (Stern 2007).

The ARPA Program has tested and demonstrated the value of public-private partnerships and different institutional models, both in implementation of the overall program and in management of individual forest sites. The program funding is disbursed through an NGO—the Brazilian Biodiversity Fund, FUNBIO—which allows greater flexibility and innovation to improve operational effectiveness and create accounts that are co-managed by protected area managers in the field for small-scale service payments and purchases. A new trust fund to finance the recurrent costs of managing these areas has been created and capitalized up to $20 million.

ARPA's innovative design has mainstreamed biodiversity conservation into land use planning and management under the Amazon's state governments and is now being replicated elsewhere. Many states are leveraging additional funds to support newly created federal and state areas. In addition, ARPA has been able to engage the private sector of Brazil and European donors to provide large funds to support protected areas. The project has worked with the WWF and many other NGOs through a collaborative and global effort to protect Amazon biodiversity. Innovative institutional arrangements are now being scaled up and replicated in other large-scale projects and programs. In late 2007, FUNBIO agreed with the state of Rio de Janeiro to develop a state environmental compensation fund and set up a program to support the state's protected areas based on the ARPA experience.

Coastal and Marine Systems as Carbon Reservoirs

Oceans are substantial reservoirs of carbon, with approximately 50 times more carbon than presently in the atmosphere (Falkowski and others 2000). They are efficient in taking up atmospheric carbon through plankton photosynthesis, mixing of atmospheric CO_2 with seawater, formation of carbonates and bicarbonates, conversion of inorganic carbon to particulate organic matter, and burial of carbon-rich particles in the deep sea. All these processes are extremely important for maintaining marine life at all tropic levels.

The current trend of increasing global atmospheric temperatures and increasing seawater acidity reduces the overall capacity of oceans to absorb more CO_2. If allowed to continue unabated, this could potentially change pH in the deep-sea regions and hinder the critical processes associated with carbon particulate burial (see box 2.8). Similarly complex relationships between water temperatures and ocean acidity in marine systems erode calcification rates in shell-bearing organisms and threaten the survival of coral reefs. Coral reefs cover less than 1 percent of the Earth's surface but are home to 25 percent of all marine biodiversity. By the end of the century, current levels of carbon dioxide emissions could result in the most acidic levels of ocean pH in 20 million years, which would have severe adverse effects on ocean water chemistry (both coastal and deep sea), the marine life and food webs, and the function of oceans as a carbon reservoir.

Coral reefs may act as a net source of atmospheric CO_2, due to the production of CO_2 during calcification. However, the fate of the free CO_2 depends on the health of the ecosystem. In healthy reefs, free CO_2 may be absorbed and recycled within the reef system. Terrestrial inputs of carbon, acidic seawater conditions, and nutrient enrichment, however, enhance the net release of CO_2 to the atmosphere. Efforts to reduce nutrient enrichment in coastal areas help to regulate ocean acidity and water temperatures, improve the quality of ocean water, and maintain

---

BOX 2.8
Crucial Role of Oceans in Climate Change

- Oceans are Earth's main buffer to climate change and will likely bear the greatest burden of impacts.
- Oceans removed about 25 percent of CO_2 emitted by human activities from 2000 to 2007.
- Oceans absorb more than 95 percent of the sun's radiation, making air temperatures tolerable for life on land.
- Oceans provide 85 percent of the water vapor in the atmosphere; these clouds are essential to regulating climate on land and sea.
- Ocean health influences the capacity of oceans to absorb carbon.

healthy corals, native fish, planktons, and seabird populations, while maintaining the carbon reservoirs (see box 2.9). Other coastal systems such as mangrove forests and sea grass beds can also be important carbon stores and sinks.

The signing of the Manado Ocean Declaration in May 2009 has made coastal and marine issues an important part of the climate change dialogue. Such issues are expected to become a central focus in future climate change negotiations (see box 2.10).

Investing in Alternative Energy

Hydropower and other sources of renewable energy such as wind and wave energy have significant potential to mitigate climate change by reducing the GHG intensity of energy production. However, large-scale hydropower development can also have high environmental and social costs, such as changes in land use, disruption of migratory pathways, and displacement of local communities. They can also disrupt environmental flows, reducing a freshwater ecosystem's potential to adapt to climate change. The ecosystem impacts of specific hydropower projects may be minimized, depending on factors such as the type and condition of pre-dam

---

BOX 2.9
The Economics for Protecting Coral Reefs

A synthesis of economic studies examining the exploitation of reefs in the Philippines demonstrated that, despite high initial benefits, destructive fishing techniques provide fewer benefits than sustainable fishing. Unsustainable fishing reduces social benefits and has a total economic value of $870 per hectare. By comparison, a healthy reef that provides tourism, coastal protection, and fisheries has a total economic value of $3,300 per hectare.

Net Present Value of Coral Reefs in the Philippines

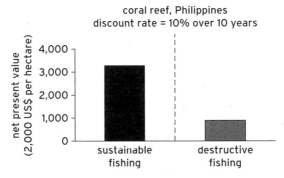

Source: Balmford and others 2002.

> **BOX 2.10**
> **The Manado Ocean Declaration**
>
> On May 14, 2009, representatives from 76 countries officially adopted the Manado Ocean Declaration at the World Ocean Conference in Indonesia. Participants recommended the inclusion of oceans and coastal areas in future climate change negotiations, including the UNFCCC Conference of Parties, to be held in December 2009 in Copenhagen. The declaration highlighted the need for (a) financial resources and incentives to assist developing countries in protecting oceans and seas, (b) renewable ocean technologies, and (c) funding for more research into the impact of climate change on oceans and the role of large bodies of water in fighting the harmful effects of climate change.
>
> The Manado Ocean Declaration emphasized the following needs:
> - Development of national strategies for sustainable management of coastal and marine ecosystems, in particular mangroves, wetlands, sea grass, estuaries, and coral reefs, as protective and productive buffer zones that deliver valuable ecosystem goods and services and have significant potential for addressing the adverse effects of climate change.
> - Cooperation in furthering marine scientific research and integrated ocean observation systems.
> - Education and public awareness to improve understanding of the role of oceans in climate change and vice versa and the role of coastal and marine ecosystems in reducing the effects of climate change.
> - Adequate measures to reduce sources of marine pollution and assure integrated management and rehabilitation of coastal ecosystems.

ecosystems, the type and operation of the dam (for example, water flow management), and the depth, area, and length of the reservoir. Run-of-the-river hydropower and small dams generally have less impact on biodiversity than large dams, but the cumulative effects of many small units should be taken into account. Careful design and planning to protect natural ecosystems in and around new facilities can benefit both biodiversity and the efficiency and effectiveness of the infrastructure investment. Protection of native forests in the watershed of the Nam Theun 2 Dam in Lao People's Democratic Republic is a critical factor in reducing soil runoff and sedimentation in the reservoir, thereby extending the lifespan of the hydropower generation facility (see box 2.11).

Construction of more dams and other irrigation infrastructure will increase due to the increasing need for alternative energy and irrigation in a warmer world. In order to maintain the mitigation and adaptation potential of freshwater ecosystems, infrastructure planning needs to take environmental flows into account. A thorough environmental flow assessment during project preparation can prevent high financial, social, reputational, and political costs. For example, in the Senegal

BOX 2.11
Nakai Nam Theun: Forest Conservation to Protect a Hydropower Investment in Lao PDR

The Nam Theun 2 Hydropower Project in central Lao PDR will inundate 450 square kilometers of the Nakai plateau, including substantial areas of seminatural forest habitat. To offset this impact, a Bank loan for the environment will provide an unprecedented level of support for conservation in the adjacent Nakai Nam Theun national protected area. At around 4,000 square kilometers (including corridors), Nakai Nam Theun is the largest single protected area in Lao PDR, with 403 species of birds and a large number of mammals, including elephants, the rare saola (*Pseudoryx nghetinhensis*), and large mammals discovered as recently as the 1990s. The protected area sits on the spine of Indochina, the Annamite Mountains, a center of high biodiversity and species endemism. The borders of Nakai Nam Theun stretch from wet evergreen forests along the Vietnamese border to the limestone karst formations of central Lao PDR, which harbor a new family of rodents that were first described in 2005. Married to this biodiversity is an astonishing ethnolinguistic diversity. The people living in, and immediately around, the protected area include 28 linguistically distinct groups who can name a greater number of forest products than have been recorded in any other area of the country.

Under a new conservation authority established during preparation of the project, the protected area will be managed according to an integrated conservation and development model. Village agreements will be developed to detail rules and regulations for zoning resource use, including controlled use and totally protected zones. Village conservation teams will provide a platform for managing natural resources and monitoring biodiversity and enforcement. Sustainable alternative livelihood options will mitigate the negative impacts resulting from restrictions on the use of resources in core conservation areas. Communities will be empowered through the provision of secure land rights, capacity building, recognition of indigenous knowledge, and equitable distribution of benefits to ensure that the most vulnerable and most forest-dependent groups are included in the process.

Previous conservation efforts in Lao PDR have been undermined by a lack of staff and long-term funding. Perhaps the most promising innovation in Nakai-Nam Theun is a new financial and administrative model. Since the protected area covers around 95 percent of the catchment for the Nam Theun 2 Hydropower Project, the developer will pay $1 million annually for protection of the area over the 30-year concession period. The Lao government is keen to apply similar financial models elsewhere, as it exploits its abundant water resources to mobilize resources for poverty reduction while maintaining the biodiversity critical for many rural households. The funding for Nakai Nam Theun will be some two orders of

(continued)

> **BOX 2.11** *(continued)*
>
> magnitude greater than the total presently allocated from the central budget to the rest of the Lao protected areas system. The Bank is therefore establishing another fund for other local conservation areas to provide modest, demand-driven funding at a level appropriate to existing local capacity. Sustained support for the fund will also come from the revenues generated by natural resource industries. Through direct financing and the promotion of integrated development models, the Bank is providing long-term biodiversity funding for conservation efforts in Lao PDR.

basin, the water charter signed by the governments of Guinea, Mali, Mauritania, and Senegal recognizes the provision of water flows to the mid-river floodplain and ensures the maintenance of agricultural and fishing activities. Similarly, the Lesotho Highlands Water Project links resource losses associated with reduced river flows to community livelihoods and downstream social impacts of the dams and offers important lessons in the following areas:

- Understanding the difference between downstream and upstream social impacts
- Recognizing the difference in magnitude in the number of people affected downstream of the dam (about 39,000 in Lesotho) compared to upstream of the dam (around 4,000)
- Developing an approach for systematically defining the affected communities (or "the population at risk") downstream of dams
- Delineating the downstream socioeconomic impacts associated with changes in river flows
- Defining approaches for addressing, and mitigating, the social impacts associated with significant changes in river flows and their limitations.

Biofuels for Renewable Energy

New initiatives under the climate change agenda provide both opportunities and challenges for biodiversity conservation. Biofuels and bioenergy plantations, for example, can substitute for fossil fuels and may also provide benefits to small farmers engaged in their production. Policies in the United States and the European Union that mandate specific targets for biofuels in meeting national fuel needs are encouraging the growth of biofuel industries. However, without careful planning, biofuel production could lead to further clearance of natural habitats, either for biofuels themselves or for new agricultural land to replace converted croplands. Moreover, many species being promoted for biofuel production are known to become invasive in some countries where they have been introduced (see table 2.2). Few current biofuel programs are economically viable without subsidies, and

TABLE 2.2
Known Invasive Species Proposed as Suitable for Biofuel Production

Species Name	Common Name	Native Range	Invasive Status
Artocarpus communis, *A. altilis*	Breadfruit	Pacific Islands, Southeast Asia	Fiji, Kiribati, Line Islands
Arundo donax	Giant reed	Eurasia	Australia, the Caribbean, Hawaii, Mexico, New Zealand, South Africa, Southern Europe, Thailand, United States
Azadirachta indica	Neem	Bangladesh, India, Myanmar, Sri Lanka,	Australia, Fiji, Mauritius, West Africa
Brassica napus	Rapeseed, canola	Eurasia	Australia, Ecuador, Fiji, Hawaii, New Caledonia
Camelina sativa	False flax	Eastern Europe and Southwest Asia	Australia, Central America, Japan, North America, South America, Western Europe
Elaeis guineensis	African oil palm	Madagascar, West Africa	Brazil, Florida, Federated States of Micronesia
Gleditsia triacanthos	Honey locust	Eastern North America	Australia, Central Argentina, New Zealand, South Africa, United States
Jatropha curcas	Jatropha, physic nut	Tropical America	Australia, Pacific Islands, Puerto Rico, South Africa, United States
Maclura pomifera	Osage orange	Central United States	Australia, Europe, South Africa, United States
Morus alba	Mulberry	Asia	Brazil, Ecuador, United States
Olea europaea	Olive tree	Mediterranean Europe	Australia, Hawaii, New Zealand
Phalaris arundinacea	Reed canary grass	Asia, Europe, North America	Australia, Chile, New Zealand, South Africa, United States, most temperate countries

(continued)

TABLE 2.2
Known Invasive Species Proposed as Suitable for Biofuel Production *(continued)*

Species Name	Common Name	Native Range	Invasive Status
Prosopis spp.	Mesquite	North America	Australia, Eastern Africa (Djibouti, Eritrea, Ethiopia, Sudan), Southern Africa, India
Ricinus communis	Castor bean	East Africa	Australia, Brazil, Mexico, New Zealand, Pacific Islands, South Africa, United States, Western Europe
Sorghum halepense	Johnson grass	Mediterranean to India	Australia, Central and South America, Indonesia, Pacific Islands, Thailand, United States
Ziziphus mauritiana	Chinese apple, jujube	China, India	Africa, Afghanistan, Australia, China, Malaysia, some Pacific archipelagoes, and Caribbean region

Source: GISP 2008.

many have potential social and environmental costs, including intensified competition for land and water and possibly deforestation. While biofuel plantations on degraded or abandoned agricultural lands may prove beneficial, the expansion of biofuels in the tropics is also leading to clearance and loss of natural ecosystems, with consequent loss of biodiversity. The clearance of peat swamp forests for oil palm production in Indonesia, for instance, is estimated to have been a major contributor to Indonesia's GHG emissions, making Indonesia the third largest emitter of GHGs in 2006.

Pilot biofuel projects of various scales are already under way or in the planning stages, particularly in Asia, Africa, and South America, to establish smallholder plantations of biofuel species, such as *Jatropha curcas*, for job creation, poverty alleviation, and restoration of degraded land. *Jatropha curcas* is a fast-growing, drought-resistant shrub or small tree that is native to southern Mexico and Central America but introduced to many tropical and subtropical countries. A member of the *Euphorbia* family, it can tolerate marginal, nutrient-poor soils and arid conditions, although it is relatively sensitive to frost. Because it is unpalatable to livestock, it has been widely used in rural communities in Africa as a hedge or "living fence" around crops. Once mature, the trees annually produce about 4 kilograms of seed, which have an oil content of 30–40 percent. The Bank is assessing the social and economic benefits of promoting *Jatropha* for biofuel production in Kenya. Biofuels may be a useful crop on degraded lands, including lands previously deforested for agricultural production, as in Brazil.

There is growing evidence however, that biofuels are not a silver bullet. Economists, environmentalists, and social scientists, among others, have presented compelling evidence that (a) some biofuels are not economically attractive alternatives to fossil fuels in the absence of subsidies; (b) they may not provide significant savings in GHG production; (c) the cultivation of plant-based biofuels may have a severe impact on biodiversity; and (d) the social impacts of the expansion of plant-based biofuels can have detrimental impacts on the poorest populations in the developing world by reducing the availability and affordability of food (see box 2.12). Accordingly, the Bank has worked with the World Wildlife Fund to produce a prototype scorecard to assess when, where, and what biofuel production is environmentally and socially sustainable. This Biofuels Sustainability Scorecard will allow the user to rate a potential biofuel on a series of criteria pertinent to the expected environmental sustainability of the biofuel and its production system.

> **BOX 2.12**
> ## Biofuels: Too Much of a Good Thing?
>
> With oil prices at record highs and with few alternative fuels for transport, several countries are actively supporting the production of liquid biofuels from agriculture—usually maize or sugarcane for ethanol and various oil crops for biodiesel. The economic, environmental, and social effects of biofuels need to be assessed carefully before extending public support to large-scale biofuel programs. Those effects depend on the type of feedstock, the production process used, and the changes in land use.
>
> Global production of ethanol as a fuel was around 40 billion liters in 2006. Of that amount, nearly 90 percent was produced in Brazil and the United States. In addition, about 6.5 billion liters of biodiesel were produced in 2006, of which 75 percent was produced in the European Union. Current biofuel policies could, according to some estimates, lead to a fivefold increase in the share of biofuels in global transport—from just over 1 percent today to around 6 percent by 2020.
>
> *Are biofuels economically viable, and what is their effect on food prices?* Governments provide substantial support to biofuels so that they can compete with gasoline and conventional diesel. Such support includes consumption incentives (fuel tax reductions), production incentives (tax incentives, loan guarantees, and direct subsidy payments), and mandatory consumption requirements.
>
> Rising prices for agricultural crops caused by demand for biofuels have come to the forefront of the debate about a potential conflict between food and fuel. Rising prices of staple crops can cause significant welfare losses for the poor, most of whom are net buyers of staple crops. But many other poor producers, who are net sellers of these crops, benefit from higher prices. For example, biofuel production has pushed up feedstock prices.
>
> *Nonmarket benefits and risks are context-specific.* The possible environmental and social benefits of biofuels are second only to energy security as the most frequently cited argument in support of public funding and policy incentives for biofuel programs. But these come with risks.
>
> *Potential environmental benefits.* Environmental benefits need to be evaluated on a case-by-case basis because they depend on the GHG emissions associated with the cultivation of feedstocks, the process of biofuel production, and the transport of biofuels to markets. Changes in land use, such as cutting forests or draining peatland to produce feedstock such as oil palm, can cancel the GHG savings for decades. Similarly, agricultural expansion arising from the need to replace land converted from food crops to biofuel production can eliminate the GHG savings and irreversibly damage wildlife and wilderness.

(continued)

Benefits to smallholders. Biofuels can benefit smallholder farmers by generating employment and increasing rural incomes, but the scope of those benefits is likely to remain limited given current technologies. Ethanol production requires fairly large economies of scale and vertical integration because of the complexity of the production process in the distilleries. Small-scale production of biodiesel could meet local demand for energy, but rising prices for food and feedstock could negate any gains in cheaper energy.

Source: World Bank 2008b.

CHAPTER 3

Ecosystem-Based Adaptation: Reducing Vulnerability

DURING THE COURSE OF HUMAN HISTORY, societies have often needed to manage the impacts of adverse weather events and climate conditions. Nevertheless, the pace of global change is now so rapid that additional measures will be required to reduce the adverse impacts of climate change in the near and long term. Moreover, vulnerability to climate change can be exacerbated by other stresses, including the loss of habitats and natural resources, reduction in ecosystem services, and land degradation.

Adaptation is becoming an increasingly important part of the development agenda, especially in developing countries most at risk from climate change. An essential component of adaptation is the protection and restoration of ecosystems and the habitats, natural resources, and services they provide. The multiple benefits of goods and services afforded by biodiversity and healthy ecosystems are largely unrecognized and unrecorded in natural accounting. Enhanced protection and management of natural ecosystems and more sustainable management of natural resources and agricultural crops can play a critical role in adaptation strategies. Ecosystem-based approaches can contribute to adaptation strategies through the following:

- Maintaining and restoring natural ecosystems and the goods and services they provide
- Protecting and enhancing vital ecosystem services, such as water flows and water quality

- Maintaining coastal barriers and natural mechanisms of flood control and pollution reduction
- Reducing land and water degradation by actively preventing, and controlling, the spread of invasive alien species
- Managing habitats that maintain nursery, feeding, and breeding grounds for fisheries, wildlife, and other species on which human populations depend
- Providing reservoirs for wild relatives of crops to increase genetic diversity and resilience.

Ecosystem-based adaptation complements other responses to climate change in two ways. First, natural ecosystems are resistant and resilient; they also provide a full range of goods and ecosystem services, including natural resources such as water, timber, and fisheries on which human livelihoods depend. Second, natural ecosystems provide proven and cost-effective protection against some of the threats that result from climate change. For example, wetlands, mangroves, oyster reefs, barrier beaches, and sand dunes protect coasts from storms and flooding. Such ecosystem-based approaches can complement, or substitute for, more expensive infrastructure investments to protect coastal settlements.

Conserving Biodiversity under Climate Change

Conservation biology confirms the need to protect large areas of habitat and maintain landscape connectivity between natural habitats and across altitudinal gradients. Many threatened and charismatic species will not survive without adequate protection of large and connected landscapes. This is especially true for wide-ranging and migratory species, such as elephants, large herbivores, and many birds, and for large carnivores at the head of the food chain. Corridors of natural habitats within transformed production landscapes or remaining habitat links between protected areas provide opportunities for species to move and maintain viable populations. Maintaining connectivity between natural habitats, and along altitudinal gradients, is a key strategy to allow plant and animal species to adapt to climate change (box 3.1).

In Colombia, a Global Environment Facility (GEF) project in the Andes has a specific component dedicated to building ecological corridors through the highly devastated cloud forests and páramo habitats of the mountain chain. More than 70 percent of Colombia's 41 million inhabitants reside in the high Andes plateaus and mountains, transforming the original habitats into agriculture and pasturelands. The project has identified new areas for conservation through private reserves and is working with farmers to raise awareness of the need to establish biological corridors.

Many Bank projects are supporting biodiversity conservation across large landscapes through improved management across mosaics of different land uses. Bank support for biological corridor projects is ensuring protection of large landscapes

BOX 3.1
Biological Corridors in a Changing World

The *Mesoamerican Biological Corridor* (MABC) is a natural corridor of tropical rain forests, pine savannas, montane forests, and coastal wetlands that extends from Mexico to Colombia. Within the corridor, the Bank is supporting national interventions in Guatemala, Honduras, Mexico, Nicaragua, and Panama to conserve the Atlantic forests of Central America. In Nicaragua, for instance, a GEF grant supported the incremental costs of protected areas and conservation-based land use in the corridor as part of an integrated development and conservation project. Management was strengthened in three key protected areas along the Caribbean coast: Cerro Silva Natural Reserve (339,400 hectares), Wawashan Natural Reserve (231,500 hectares), and the Cayos Miskitos Biological Reserve, which protects nesting grounds of five of the world's seven species of marine turtles. Within the corridor, indigenous communities were assisted to gain tenure over indigenous lands and to develop livelihoods based on sustainable management of natural habitats and resources. Recent studies have shown that the MABC forests are significant areas of high carbon storage.

The *Atlantic Forest of Brazil* is one of the most threatened ecosystems in Latin America, where only 7 percent of the original habitat remains in a few isolated forest patches. The area has an extraordinarily high level of endemism. The Bank, through the Pilot Program to Conserve the Brazilian Rain Forest and G-8 donors, is working to improve the connectivity of these patches through an ecological corridors project, which brings together states, municipalities, NGOs, and academic institutions. Similarly, in the highly threatened Chaco Andean system in Ecuador, a Bank-funded project has strengthened biological corridors through funding for private reserves and innovative conservation models.

The *Critical Ecosystem Partnership Fund* (CEPF) is supporting civil society activities to address threats to biodiversity across landscapes that include a matrix of uses, from protected areas to high-value conservation sites in production landscapes. A critical ecosystem profile identifies the priorities for each hot spot. Many of those high-priority activities are targeted toward key biological corridors, which overlay areas of high carbon. CEPF has supported activities in Sierra Madre in the Philippines, Barisan Selatan in Sumatra, key forest corridors in Madagascar, the West Guinea and Eastern Arc forests in Africa, mountain corridors in the Caucasus and eastern Himalayas, and the Choco-Manabi and Vilcabamba-Amboro corridors in the tropical Andes. A new phase of funding will target important biological landscapes in Indochina, including the Mekong corridor, and the highly diverse tropical forests of the Western Ghats in India.

and biological corridors, promoting connectivity in the Maloti-Drakensberg transfrontier region in Lesotho and South Africa; mega-reserves from mountains to the sea in the Cape region; corridors in the Vilicabamba-Amboró region in República Bolivariana de Venezuela, Colombia, Ecuador, Peru, Bolivia, and northern Argentina; and a network of corridors in Bhutan. Transboundary conservation efforts in the West Tien Shan in Central Asia foster international collaboration and cooperation across national boundaries, reduce disturbance on fragile mountain grasslands, and promote conservation of wide-ranging species. A new Tien Shan ecosystem development project will promote further protection for the juniper and walnut forests and other key mountain habitats. The project covers the Kyrgyz Republic and Kazakhstan and benefits from funds through the GEF and the BioCarbon Fund in recognition of the important role that mountain ecosystems play in regulating ecosystem services and carbon sequestration.

Maintaining and Restoring Natural Ecosystems

Within any given ecosystem, functionally diverse communities are more likely to be resilient to climate change and climate variability than biologically impoverished communities. Habitat conservation and protected areas play an important and cost-effective role in protecting biological resources and reducing vulnerability to climate change. The Bank recognizes the important role that enhanced protection of natural forests can play in protecting development investments. Thus the Dumoga-Bone National Park in Indonesia was established to protect a major irrigation investment in northern Sulawesi. Similarly, a new conservation area in Lao People's Democratic Republic protects the forests around the Nam Theun 2 Dam and its watersheds (see box 2.11), reducing sedimentation in the reservoirs and extending the lifespan of the hydropower generation facility. Coastal protected areas in Croatia, Bangladesh, Indonesia, Honduras, and Lithuania are protecting coastal forests, swamps, floodplains, and mangroves, which are important for shelter belts and flood control. The role of natural habitats in providing services such as coastal protection and nursery grounds for quality fisheries is increasingly being recognized as essential to these countries' coastal economies and the livelihoods of the communities who depend on them.

Improved management of natural habitats and the reduction of threats such as habitat conversion, overharvesting, pollution, and alien species invasions contribute to healthier and more resilient ecosystems. For example, reducing the pressures from coastal pollution, overexploitation, and destructive fishing practices improves the health of coral reefs and increases their resilience to rises in water temperature and bleaching. Similarly, countering habitat fragmentation through the protection or establishment of biological corridors between protected areas increases the resilience of forests. More generally, mosaics of interconnected terrestrial, freshwater, and marine multiple-use areas and protected reserves are better adapted than fragmented habitats to meet conservation and livelihood needs under changing climate conditions. Such ecosystem-based approaches are low-cost, long-proven, and low-technology solutions to many of the anticipated adverse impacts arising from climate change.

Wetlands are some of the most threatened ecosystems on Earth, yet they provide many vital ecosystem functions. Montane wetlands and freshwater rivers and lakes serve as vital water recharge areas and important sources of water for irrigation and domestic and industrial use. Freshwater and coastal wetlands downstream are also productive fisheries on which many of the world's poorest communities depend. Wetlands can also act as filters, removing pollutants and improving water quality. In Bulgaria, the Bank is working with the World Wide Fund for Nature (WWF) and other partners to restore natural wetlands along the Danube River as filter beds to remove pollutants and provide habitat for native wildlife (see box 3.2).

As climate change exacerbates the impacts of environmental stresses, many of the free goods and services provided by natural habitats will become ever more valuable. Enhanced protection of natural wetlands and, increasingly, restoration of wetland habitats will become an important adaptation strategy (see box 3.3).

Reducing Vulnerability

Natural ecosystems can also reduce vulnerability to natural hazards and extreme climatic events. Protecting forests and other natural ecosystems can provide social, economic, and environmental benefits, both directly through more sustainable management of biological resources and indirectly through protection of ecosystem services. Mountain habitats, for instance, bestow multiple ecosystem, soil conservation, and watershed benefits. They are often centers of endemism, Pleistocene refuges, and source populations for restocking of more low-lying habitats. Mountain ecosystems influence rainfall regimes and climate at local and regional levels, helping to contain global warming through carbon sequestration and storage in soils and plant biomass. Wetlands are nature's kidneys, providing indispensable ecosystem services that regulate nutrient loading and water quality.

Over the last decade, more and more Bank projects have been making explicit linkages between sustainable use of natural ecosystems, biodiversity conservation, carbon sequestration, and watershed values associated with erosion control, clean water supplies, and flood control. Bank watershed projects in the Middle East incorporate natural forests and endemic riparian woodlands as part of micro-catchment vegetation management working with local communities in the Lakhdar watershed in Morocco, the wadis in the northern Republic of Yemen, and the eastern Anatolia Basin in Turkey. In China, mountain forests are being increasingly recognized for their role in the supply of clean water, regulation of water flows, and control of floods. The China Forest Protection Project is focusing on mountain and upper

> **BOX 3.2**
> ## Restoring the Lower Danube Wetlands
>
> Conversion of floodplains for farming and other development has led to 95 percent of the Upper Danube, 75 percent of the Lower Danube, and 28 percent of the delta's floodplains being cut off by dikes. This has increased the risk of floods and pollution in the region, threats that are expected to rise with climate change.
>
> In 2000 WWF secured agreement from the heads of state of Bulgaria, Romania, Moldova, and Ukraine to restore 2,236 square kilometers of floodplain to form a 9,000 kilometer Lower Danube Green Corridor. This corridor is intended to attenuate floods, restore biodiversity, improve water quality, and enhance local livelihoods. As of 2008, 469 square kilometers of floodplain, 14.4 percent of the area pledged, has been or is undergoing restoration. In Romania, the Babina and Cernovca polders have been reflooded, and in Ukraine, the Tataru polder has been flooded to link the Katlabuh Lake to the river. Restoration of the pilot polders has seen a diversification in livelihood strategies to fishing, tourism, reed harvesting, and livestock grazing on seasonal pastures, activities that earn an average of €40 ($56) per hectare per year. At Babina and Cernovca polders, the restored fisheries provide jobs for 20-25 people.
>
> Restoration activities at Katlabuh Lake have improved water quality for 10,000 local residents. The value of ecosystem services, such as restored floodplains for fisheries, forestry, animal feed, nutrient retention, and recreation, is estimated at €500 ($698) per hectare per year, or around €85.6 million ($119 million) per year for the restoration area. Following restoration, the number of resident breeding bird species increased from 34 to 72. As a result of its accession to the European Union, Romania has designated an additional 5,757 square kilometers as Natura 2000 protected areas. Restoration of the 37 sites that make up the Lower Danube Green Corridor will cost an estimated €183 million ($299 million), but will likely generate additional earnings of €85.6 million ($120 million) per year. Before the restoration, the 2005 flood cost €396 million in damages, proving the cost-effectiveness of ecosystem-based approaches.
>
> *Source:* WWF 2008.

watershed forests and redesignating forests for their watershed and biodiversity protection functions as well as for more sustainably managed production.

The Bank has been a leader in piloting payments for ecosystem services (PES). In Mexico, Bank projects have helped to establish PES systems to reduce logging in the Monarch Butterfly Reserve in an effort to protect an important butterfly habitat. With support from the Mexican Nature Conservation Fund, an endowment has been established for El Triunfo Reserve in the Sierra Madre in Chiapas to support activities that protect the area's ecosystem services, especially water

> **BOX 3.3**
> ## Rebuilding Resilience in Wetland Ecosystems
>
> The Gulf of Mexico possesses one of the richest, most extensive, and most productive ecosystems on Earth—coastal wetlands that cover an area of more than 14,000 square kilometers. The coast is flanked by 27 major systems of estuaries, bays, and coastal lagoons that serve as shelter, feeding, and breeding areas for numerous species of important riverine and marine fishes. Moreover, the coastal swamps of Campeche and Tabasco are home to 45 of the 111 endemic species of aquatic plants in Mexico. These coastal wetlands play an important role in the water cycle.
>
> Climate change is already having an impact on these ecosystems. Sea-level rise in the Gulf of Mexico is leading to saltwater intrusion. Anticipated modifications in rainfall patterns in northern Mexico will affect natural drainage systems, further deteriorating the natural water balance of these coastal wetland systems. Degraded marshlands and mangroves will be less likely to withstand extreme weather events. The number of high-intensity hurricanes that have reached landfall in the Gulf of Mexico has increased by more than 40 percent compared to the 1960s. These storms often cause serious disruption, with loss of property and human life. The ecological and economic consequences can be staggering.
>
> The Bank is preparing a project to address these concerns through improved water resource and wetland management. The project will pilot several measures:
>
> - Restoring wetlands, taking into account sand dynamics and hydrology (initial activities will include the removal of soil or sand sediments obstructing water flows and the maintenance of waterways that feed wetland restoration)
> - Integrating climate change adaptation measures into resource management programs
> - Restoring mangrove swamp ecosystems by establishing permanent and seasonal closed areas as well as by reducing and preventing changes in land use, promoting more efficient water management, and reintroducing native mangrove species in areas degraded by economic activities
> - Maintaining water supply for production sectors
> - Developing mechanisms to promote sustainable land use patterns that maintain the functional integrity of wetland ecosystems in the region.

production. In Ecuador, an integrated watershed management project is being prepared with a specific component to capture payment for environmental services provided by Andean forests. Meanwhile, Costa Rica is launching a second Bank-GEF project to build on the experience and success of the Ecomarkets Projects in promoting biodiversity conservation and PES schemes on privately owned lands (see box 3.4).

> **BOX 3.4**
> ## Ecomarkets in Costa Rica
>
> Costa Rica's Program of Payments for Environmental Services (PSA) is an innovative and highly successful effort to enlist private landholders to maintain and protect their forests voluntarily. Since its inception in 1997, the PSA Program has been applied to nearly 500,000 hectares of privately owned forests.
>
> Since 2001, the program has received funding under the Bank-GEF Ecomarkets Project. More than 130,000 hectares of high-priority biodiversity areas in the Costa Rican portion of the Mesoamerican Biological Corridor have been included in the program. Another 70,000 hectares have been contracted on privately owned lands within other high-priority conservation areas, further contributing to the achievement of conservation and sustainable management goals. In 2000, only 22 female landholders participated in the program; by 2005, 474 women were participating. In 2000, 2,850 hectares of indigenous community-owned lands were in the program; by 2005 this figure had risen to 25,125 hectares, an eightfold increase.
>
> The PSA Program has been funded primarily by allocating 3.5 percent of the national fuel tax to the Fondo Nacional de Financiamiento Forestal (FONAFIFO). It has also attracted significant co-financing from bilateral donors, including Germany, Norway, and Japan. The Ecomarkets Project has not only provided additional financing to expand the program but also refocused it on global and regional biodiversity conservation as well as on national social goals. National benefits include the maintenance of privately owned forests in important biological corridors; local conservation of biological diversity; increased involvement of women landholders and indigenous communities with the PSA Program; direct payments to a greater number of small rural landholders; and, most important, broad-scale public recognition that intact forests and their environmental services have value.
>
> The success of the Ecomarkets Project is based on a strong institution (FONAFIFO) that is capable of effectively and efficiently managing a complex system of payments for environmental services; the strong legal framework and wide political support for the PSA Program through three successive administrations; and the nationwide support from civil society, particularly small- and medium-size landholders as well as local and regional organizations (for example, NGOs and cooperatives). The PSA Program and the Ecomarkets Project have attracted widespread international interest, spurring several replication efforts. FONAFIFO has hosted official delegations from many countries wanting to study the program. The project has led to more effective conservation by creating linkages between geographically isolated protected areas through privately owned lands where biodiversity is legally protected through PSA contracts.

Adopting Indigenous Knowledge to Adapt to Climate Change

Indigenous peoples can play a key role in mitigating and adapting to climate change. Many territories of indigenous groups have been better conserved than the adjacent agricultural lands, including in Brazil, Colombia, and Nicaragua. As satellite maps clearly show, the area of the Amazon covered by indigenous lands represents one of the largest remaining reserves of intact tropical forest. These indigenous groups are in a good position to participate in the various private and public carbon payments for avoided deforestation. A climate change agenda fully involving indigenous peoples has many more benefits than one involving only government or the private sector. Indigenous peoples are especially vulnerable to the negative effects of climate change, but they are also a source of knowledge and adaptation strategies. For example, ancestral territories often provide excellent examples of a landscape design that can resist the negative effects of climate change. Over the millennia, indigenous peoples have cultivated genetic varieties of medicinal and useful plants and animal breeds with a wider natural range of resistance to climatic and ecological variability. They also have evolved farming and water management strategies to cope with climate change (see box 3.5).

Over the last two decades, 109 Bank projects have supported, or are supporting, indigenous peoples' programs and needs. Several of these projects have supported the conservation of tropical forests and reforestation activities linked directly to avoided deforestation; a few have supported direct benefits from carbon payments. The following activities supporting climate change and indigenous objectives are common components of these projects: (a) establishment of indigenous reserves and co-management of protected areas, (b) titling and demarcation of indigenous lands, (c) indigenous life plans, (d) indigenous community management and zoning plans, (e) indigenous community mapping and conservation, (f) community sustainable livelihoods, and (g) capacity building and training.

Adaptation in Coastal Areas

Coastal wetlands act as natural barriers, protecting coastal settlements from storms and other natural hazards and reducing the risk of disaster. Mangroves and other coastal wetlands are especially vulnerable to climate change and rising sea levels. The loss of mangroves, in turn, makes coastal communities vulnerable to extreme events such as hurricanes, cyclones, and tsunamis. Inland areas protected by healthy mangroves have generally suffered less than more exposed communities from extreme weather events such as Cyclone Nargis, which hit southern Myanmar in 2008, and the destructive tsunami that hit Southeast Asia in 2004. As well as providing coastal defenses, mangroves are important nurseries for fish, prawns, and other marine invertebrates that are critical resources for local livelihoods.

> **BOX 3.5**
> **Measures to Address Climate Change in the Salinas and Aguada Blanca National Reserve in Peru**
>
> Since 2005 the GEF has supported the Participatory Management of Protected Areas Project in Peru, including the Salinas and Aguada Blanca National Reserve. Located north of the city of Arequipa, at an altitude between 3,600 and 6,000 meters, the Salinas and Aguada Blanca National Reserve is home to wild cameloids, such as vicuña and guanaco, as well as many migratory and resident birds that breed around the mountain lakes, dams, and rivers. Created in 1979 to conserve the endangered flora and fauna of the area, the reserve recently has been extended to 366,936 hectares. The volcanoes Misti, Chachani, and Pichu Pichu lie within the reserve, as does the beautiful Salinas lagoon, which creates an ideal habitat for flamingos.
>
> The reserve protects the main source of water that supplies the city of Arequipa as well as smaller towns. The natural ecosystems are threatened from deforestation by the 8,000 inhabitants from 14 local communities living within the reserve, many of whom are engaged in cameloid farming. Water resources are becoming increasingly scarce due to the melting of the glaciers and because the area receives less precipitation than in the past, a decline that can be attributed to climate change. The GEF project has supported subprojects to help the local communities adapt to climate change, including water conservation and management activities that have also contributed to biodiversity conservation. The project supports terracing to collect water during the rainy season and measures to improve infiltration and water conservation. It has reintroduced technology developed and used by the indigenous inhabitants before the Spanish Conquest, including infiltration ditches, small barrages, water mirrors (small lakes), and rustic canals. After a few years of implementation, water availability has improved, especially during the summer season, and vegetation has recovered in some parts of the reserve.

Restoration of degraded mangroves in the Mekong Delta in Vietnam, for example, has improved management of coastal forests, safeguarding important nursery grounds for local fisheries and food security (see box 3.6).

Rising sea levels cause significant change to ecosystems and loss of marine resources. The construction of dikes and seawalls, as well as other coastal development and infrastructure, may further degrade natural habitats and increase the stress on coastal resources. Small-island states are especially vulnerable to climate change. Accordingly, some of the first Bank projects on adaptation focused on small-island states in the Pacific (Kiribas) and the Caribbean. Caribbean Planning for Adaptation to Climate Change (CPACC), a regional activity, focused on the vulnerability of the island nations of the Caribbean to the impacts of climate

> **BOX 3.6**
> **Investing in Mangroves**
>
> The destruction of mangroves has a strong economic impact on local fisheries and fishing communities. Maintenance or restoration of mangroves can, however, reduce vulnerability of coastal areas to sea-level rise and extreme weather events, while also contributing to food security. Often such ecosystem-based approaches are highly cost-effective.
>
> Restoring and protecting mangroves can reduce vulnerability in various ways:
>
> - Mangrove forests have an estimated economic value of $300,000 per kilometer as coastal defenses in Malaysia (Ramsar Convention on Wetlands 2005).
> - Since 1994, communities have been planting and protecting mangrove forests in Vietnam as a way to buffer against storms. An initial investment of $1.1 million saved an estimated $7.3 million a year in sea dike maintenance and significantly reduced the loss of life and property from Typhoon Wukong in 2000 in comparison with other areas (IFRC 2002).
> - Loss of mangrove area has been estimated to increase storm damage on the coast of Thailand by $585,000 or $187,898 per square kilometer (in 1996 U.S. dollars), based on data from 1979–96 and 1996–2004, respectively (Stolton, Dudley, and Randall 2008).
> - Recent studies in the Gulf of Mexico suggest that mangrove-related fish and crab species account for 32 percent of the small-scale fisheries in the region and that mangrove zones can be valued at $37,500 per hectare annually (Aburto-Oropeza and others 2008).
> - In Surat Thani, Thailand, the sum of all measured goods and services of intact mangroves ($60,400) exceeds that of shrimp farming from aquaculture by around 70 percent Balmford and others 2002).

change. Potential economic impacts of climate change for the Caribbean Community (CARICOM) countries, for instance, are estimated at between $1.4 billion and $9 billion, assuming no adaptation measures. The largest category of impacts is the loss of land, housing, other buildings, and infrastructure due to sea-level rise. Impacts on agriculture are also potentially significant for CARICOM countries. According to Vergara (2005), most of the remaining impacts are due to reduction in tourism, caused by rising temperatures and loss of beaches, coral reefs, and other ecosystems (15–20 percent), and damage to property and life, caused by the increased intensity of hurricanes and tropical storms (7–11 percent).

CPACC has provided information on the bleaching of corals caused by exposure to high temperatures and explored the ecological and economic consequences for the economies of the Caribbean through monitoring stations in the Bahamas,

Belize, and Jamaica. Project data confirm the deteriorating state of coral reefs in the Caribbean and the need to create marine protected areas. Similarly, the global Coral Reef Targeted Research Project is providing the scientific underpinning for reef and fisheries management to address the threats arising from global warming (see box 3.7). Regional working groups have been established to monitor coral reefs, investigate the impacts of climate change, and design appropriate management responses.

BOX 3.7
Addressing the Impacts of Climate Change on Ocean Ecosystems and Coastal Communities

The International Year of the Reef 2008 saw a worldwide campaign to raise awareness about the value and importance of coral reefs and the need to protect them. Threats to reefs include climate change, which is leading to widespread coral damage. An unprecedented climatic event affected the world's oceans in 1998, when a strong El Niño–Southern Oscillation episode caused abnormally high sea-surface temperatures and affected more than 16 percent of the world's coral reefs. This event emphasized the urgent need to protect natural resources and to prepare coastal-dependent people to adapt to climate change. At the same time, human population growth in tropical coastal zones is exerting tremendous pressure that degrades and threatens coral reefs and associated resources.

The Coral Reef Targeted Research and Capacity Building for Management (CRTR) Program is a proactive research and capacity-building partnership designed to improve the scientific knowledge needed to strengthen management and policy to protect coral reefs. The CRTR is filling crucial gaps in targeted research areas such as coral bleaching, connectivity, coral diseases, coral restoration and remediation, remote sensing and modeling, and decision support. The CRTR partnership was formed to build national capacity for management-driven research and to use this information to improve the management of coral reefs and the welfare of the human communities that depend on them. The program is working with stakeholders and local governments through its regional centers of excellence to increase awareness of the growing risks facing coral reefs from local and global sources and the economic and social implications for the tens of millions of people who depend on them for livelihoods, food security, and coastal protection.

While policy makers in the international arena grapple with formulas and cost-effective means to bring down CO_2 emissions to well below 1990 levels over the next 50 years, the CRTR is helping local marine resource managers to buy time for coral reefs. A number of interventions are addressing immediate threats to reef ecosystems and seeking to increase their resilience to changing ocean conditions.

Fish form the primary source of protein for nearly 1 billion people and constitute a significant part of the diet for many more. Rising demand for food has left half of wild marine fisheries fully exploited, with a further quarter overexploited. Apart from the direct effects of overfishing, fish populations are threatened by higher ocean temperatures, lower water flows, changing salinity, changing seasonality of stream flow, loss of habitat, and declining water quality. Overfishing changes the structure of the food web; for example, jellyfish have replaced fish as the dominant planktivores in some waters around the United Kingdom, and there is some concern that these shifts may be difficult to reverse, since jellyfish eat the eggs of their fish competitors. There is growing evidence that species diversity is important for marine fisheries, both in the short term, by increasing productivity, and in the long term, by increasing resilience.

Marine Protected Areas

Like terrestrial protected areas, marine protected areas are created to achieve long-term biodiversity conservation but may also maintain coastal and marine resources, sustain fisheries, and provide opportunities for recreation, tourism, and research. Approximately 5,000 marine protected areas globally cover about 2.2 million square kilometers of the marine environment (Laffoley 2008). When effectively designed and managed, marine protected areas can deliver many ecological and socioeconomic benefits as well as mitigate the effects of increasing carbon emissions.

The Protected Area Program of Work of the Convention on Biological Diversity (CBD) has emphasized the need to expand the global network of marine protected areas to protect 10 percent of marine habitats by 2012. Such a network, especially when it links source and sink areas, can help to maintain functional marine ecosystems and mitigate carbon fluxes. All marine habitats are under-represented in marine protected areas, but there is a particular need to ensure that protection is extended to offshore and deep-sea areas as well as coastal reserves. The high seas beyond the legal jurisdiction of nations cover nearly 50 percent of Earth's surface but account for 90 percent of the planet's biomass (Corrigan and Kershaw 2008). Ecosystems in the high seas provide valuable functions and services, including carbon sequestration and storage and access to scientific research, exploration, and tourism.

Bank-led sector work on marine management has determined that marine protected areas provide many benefits to marine conservation, fisheries stocks, and carbon sequestration:
- Increasing the density, biomass, individual size, and diversity of all fish functional groups in communities ranging from tropical coral reefs to temperate kelp forests

- Achieving up to a 20–30 percent increase in the diversity and average size of fish in marine protected areas relative to unprotected areas
- Conserving fish populations and their habitats, thereby enhancing the marine carbon sink
- Reducing the need for engineered structural defenses, which neither provide ecosystem services nor sequester carbon.

The Bank has invested in a diverse range of marine conservation and resource management projects. Programs such as the Mesoamerican Barrier Reef Project and the Coral Reef Rehabilitation and Management Project (COREMAP) in Indonesia have recognized the important links between sources and sinks and are helping to protect some of the world's most biodiverse coral reefs through strengthened protection and community engagement in resource management (see box 3.8). Similarly, the Namibian Coast Conservation and Management Project aims to mainstream biodiversity conservation into sectoral policies and programs by providing defined incentives to stakeholders. Elsewhere, projects in Central America, Tanzania, and Vietnam have focused on integrating coastal zone management with enhanced protection of mangroves, coastal wetlands, and offshore reefs that sustain local fisheries and thriving tourism industries.

BOX 3.8
COREMAP: Coral Reef Rehabilitation and Management Project in Indonesia

The Indonesian archipelago is a center of coral and marine diversity with some of the most species-rich reef ecosystems in the world. The fisheries they support are an important source of food and economic opportunities for about 67,500 coastal villages, but reef fisheries have been increasingly threatened and overexploited in the last decade. In 1988, the government of Indonesia initiated a multidonor Coral Reef Rehabilitation and Management Program, as a 15-year national program over three phases. As one of the main donors, the World Bank financed efforts to improve the management of coral reef ecosystems in several pilot sites, including the marine protected area at Taka Bone Rate, the world's third largest atoll. Other pilot efforts in the Padeido Islands, Papua, and Nusa Tenggara supported community management of coral reefs.

The first phase of COREMAP highlighted some of the challenges facing coral reefs and the communities who depend on them. Many of the coral reef ecosystems in Indonesia, and the small-scale fisheries they support, have reached a level and mode of exploitation whereby the only way to increase future production and local incomes is to protect critical habitats and reduce fishing effort. A growing body of empirical evidence suggests that marine reserves can rejuvenate depleted fish stocks in a matter of years when they

These Bank efforts in coastal and marine management are complemented by new engagement in partnerships with other donors and major nongovernmental organizations (NGOs). The Coral Triangle Initiative in the Indo-Pacific region, for instance, aims to balance coastal protection and biodiversity conservation with improved fisheries management and local livelihoods (see box 3.9). Such projects help to maintain healthy oceans by promoting the protection and sustainable management of biologically diverse ecosystems.

Investing in Ecosystems versus Infrastructure

Adaptation to climate change is likely to involve more investment in dams and reservoirs to buffer against increased variability in rainfall and runoff. Investments in water resource infrastructure, especially dams for storage, flood control, or regulation, may be essential for economic development, enabling hydropower generation, food security and irrigation, industrial and urban water supply, and flood and drought mitigation. Nevertheless, traditional engineered solutions may work against nature, especially when they lead to loss of habitat, are poorly planned, designed, or operated, or cause problems for downstream ecosystems and communities

---

are managed collaboratively with the resource users and form the core of a wider multiple-use marine protected area. For the second phase of COREMAP, the government of Indonesia has made an important policy shift toward promoting marine conservation and protected areas to achieve sustainable management of coral reef ecosystems and small-scale fisheries.

COREMAP II will help to establish marine reserves through a participatory planning process with communities, with the goal of rejuvenating coral reefs and reef fisheries. This six-year $80 million program is being implemented in 12 coastal districts, including 1,500 coastal villages with more than 500,000 residents. The centerpiece of these efforts will be the creation of collaboratively managed marine reserves, many within existing national parks and marine protected areas of recognized global value. The government of Indonesia has committed to setting aside 30 percent of the total area of coral reefs in each participating district as collaboratively managed and fully protected marine reserves by 2030. A key component of the program is a learning network linking key marine sites and conservation efforts throughout the archipelago to exchange lessons learned and expertise. This ambitious program places Indonesia as a global leader in marine and coral reef conservation efforts. These lessons will be integrated into capacity-building efforts under the Coral Triangle Initiative to prepare local governments and communities to manage coral reefs and their associated ecosystems.

> **BOX 3.9**
> **Coral Triangle Initiative on Coral Reefs, Fisheries, and Food Security**
>
> The Coral Triangle region in the Indo-Pacific Ocean is a global hot spot for marine life and covers the economic zones of Indonesia, Papua New Guinea, Malaysia, Philippines, Timor Leste, and the Solomon Islands. Destructive fishing practices and overexploitation of coastal and marine fisheries for local and export markets are leading to a loss of marine resources that is affecting the welfare of the coastal population, who are heavily dependent on the sea and its resources for livelihoods. The loss of marine resources could affect more than 200 million people. In response, the governments of the six countries have launched the Coral Triangle Initiative.
>
> The initiative is a multidonor effort centered on high-level political commitment and proactive management by governments of the Coral Triangle area and supported by the private sector and civil society partners. It seeks to safeguard the region's marine and coastal biological resources for the sustainable growth and prosperity of current and future generations.
>
> The initiative's main objective is to advance integrated ecosystem-based management of ocean and coastal areas at regional and national levels through coordinated planning that builds on lessons gleaned from management of large marine ecosystems, marine protected areas, and community management. As part of the initiative, two large marine protected areas will be established in the Sulu Sea of Indonesia and in the Kimbe Bay-Bismarck Sea of Papua New Guinea. A network of smaller marine protected areas will combine science-based marine protection with community-based measures matching the socioeconomic needs of the local populations. Lessons learned in the process of designing and implementing the network of protected sites will be applicable to other large marine ecosystems and will benefit local communities through improved conservation of coral reefs and more sustainable fisheries.

through their impacts on the volume, pattern, and quality of flow (Hirji and Davis 2009). Instead, in Argentina and Ecuador, flood control projects use the natural storage and recharge properties of critical forests and wetlands by integrating them into "living with floods" strategies that incorporate forest protected areas and riparian corridors (see box 3.10).

Strengthening the protection of cave systems and natural forests can safeguard important aquifers and freshwater supplies. For example, the value of the Lužnice floodplain in the Czech Republic—one of the last floodplains with an unaltered hydrological regime—is quantified at $27,068 per hectare, because of a range of ecosystem services including flood mitigation, water retention, and carbon sequestration. Similarly, the value of forests for preventing avalanches is estimated at

> BOX 3.10
> ## Protecting Natural Forests for Flood Control
>
> The irregular rainfall patterns prevailing in Argentina cause floods and droughts. Under all climate change scenarios, these boom-and-bust cycles will be exaggerated. Currently, about one-fourth of the country is repeatedly flooded. This is particularly true for northeastern Argentina, which has three major rivers—the Paraná, the Paraguay, and the Uruguay—and extensive, low-lying plains. The seven provinces of this area (Entre Ríos, Formosa, Chaco, Corrientes, Misiones, Buenos Aires, and Santa Fe) make up almost 30 percent of the country and include more than half of Argentina's population.
>
> Flooding is the major force regulating the ecosystems around these rivers; virtually all ecological events in the floodplains are related to the extent and regularity of flooding. Typical habitats include the Pampas grasslands, Mesopotamia savanna, Paraná forests, Chaco estuaries and forests, and the Paraná River islands and delta. The Paraná forests in the province of Misiones have the highest level of faunal biodiversity, followed by the Chaco estuaries and forests. Overall, 60 percent of Argentina's birds and more than 50 percent of its amphibians, reptiles, and mammals are found in the floodplains.
>
> The first phase of a two-stage flood protection program provided cost-effective flood protection for the most important economic and ecological areas and developed a strategy to cope with recurrent floods. Activities included the development and enforcement of flood defense strategies, the maintenance of flood defense installations, early flood warning systems, environmental guidelines for flood-prone areas, and flood emergency plans. Extensive areas of natural forest were protected as part of the flood defense system. This incorporation of natural habitats into flood defenses provided a low-cost alternative to costly infrastructure, with the added benefit of high biodiversity gains. As changing climate makes extreme weather events and flooding more likely, the experience of Argentina provides some useful lessons on how best to harness natural habitats to reduce the vulnerability of downstream communities.

around $100 per hectare per year in open lands in the Swiss Alps and at more than $170,000 per hectare per year in built-up areas (ProAct Network 2008). Elsewhere, improved ecosystem management can reduce vulnerability and protect against natural disasters (see table 3.1).

The Bank has valuable experience integrating protection and improved management of natural ecosystems into infrastructure projects as part of sustainable development. Such projects have gone beyond measures to mitigate environmental impacts by including natural forests as part of overall flood abatement, irrigation, and coastal defense measures (see table 3.2).

TABLE 3.1
Ecosystem-Based Approaches to Defend against Natural Disasters

Natural Hazard	Types of Ecological Protection	Examples
Flood	Dense vegetation cover within upper watershed areas increases infiltration of rainfall and reduces surface runoff, reducing peak flow rates except when soils are fully saturated. Vegetation also protects against erosion, reducing soil loss and transport of mud and rock, which greatly increases the destructive power of floodwaters.	Hurricane Jeanne hit several Caribbean islands, but the number of flood-related deaths was more than 3,000 in Haiti compared with a few dozen in all other affected countries, due in large part to Haiti's highly degraded and flood-responsive watersheds. The pattern of economic losses was similar during the 2008 hurricane season, although the loss of life was far lower.
	Dense vegetation protects riverbanks and adjacent land and structures from erosion by floodwaters.	A study around Mantadia National Park, Madagascar, concluded that conversion from primary forest to swidden can increase downstream storm flow by as much as 4.5 times.
	Wetlands and floodplain soils absorb water, reducing peak flow rates downstream.	Communities have planted bamboo to protect channel embankments from annual floods in Assam. Canalization and drainage in the Mississippi floodplain reduced flood storage capacity by 80 percent and have been linked to subsidence of large areas and the severity of the impact from Hurricane Katrina.
Tsunami, storm surge	Coral reefs and sand dunes (which in coastal areas typically depend on associated plant communities for maintenance) provide a physical barrier against waves and currents.	Modeling for the Seychelles suggests that wave energy has doubled partially as a result of changes in the structure (due to bleaching) and species composition of coral reefs. In the Caribbean, more than 15,000 kilometers of shoreline could experience a 10–20 percent reduction in protection from waves and storms by 2050 as a result of reef degradation.
	Salt marshes and lagoons can divert and contain floodwaters.	Re-establishment of salt marshes forms part of coastal defense measures in the United Kingdom.
	Mangroves and other coastal forests can absorb wave energy and trap floating debris, reducing the destructive power of waves.	Data from two villages in Sri Lanka that were hit by the devastating Asian tsunami in 2004 show that, while two people died in the settlement with dense mangrove and scrub forest, up to 6,000 people died in the village without similar vegetation. In Japan, where good historical records exist, the role of forests in limiting the effects of tsunami damage have been demonstrated.

(continued)

TABLE 3.1
Ecosystem-Based Approaches to Defend against Natural Disasters (continued)

Natural Hazard	Types of Ecological Protection	Examples
Landslide	Dense and deep-rooted vegetation helps to bind soil together, resisting slippage of surface layers.	China's Grain for Green Program bans logging and agriculture on steep slopes and prohibits forest clearing for shifting agriculture in the mountains of southwestern China. In exchange, the local communities receive grain and cash subsidies as well as resilience against flooding events.
Avalanche	Forests form a physical barrier against avalanches and pin down the snowpack, reducing the chance of a slide.	Reforestation has been used for avalanche protection in Switzerland, complementing and in some cases substituting for engineered barriers.

Source: World Bank forthcoming.

TABLE 3.2
Exploring the Impacts and Offsets of Infrastructure Projects to Protect Carbon Sinks and Ecosystem Services

Sector	Environmental Impacts	Mitigation and Conservation Actions
Energy, hydropower	Flooding of natural habitats near reservoirs; displacement and loss of wildlife; loss of biodiversity; deterioration of water quality; accumulation of vegetation before reservoir filling; upstream and downstream hydrological changes; alteration of fish communities and other aquatic life; invasion of aquatic vegetation and its associated disease vector species; sedimentation of reservoirs; generation of quarries and borrow pits; construction of multiple dams in one river; human resettlement; changes in hydrodynamics	Creation of compensatory protected areas; species conservation in situ and ex situ; minimization of flooded habitats; water pollution control and vegetation removal; water release management; minimum (ecological) stream flow maintenance year round; construction of fish passages and hatchery facilities; application of fishing regulations; physical removal of containments; biological and mechanical pest control; drawdown of reservoir water levels; watershed management; sediment management techniques; landscape treatment; environmental assessment of cumulative impacts
Energy (pipelines), transportation (roads), telecommunications (access corridors)	Barriers to species dispersal; habitat loss, fragmentation, and simplification; spread of tree diseases; insect infestation; introduction of invasive species; human and domestic animal intrusions; runoff, erosion, and landslides; fire generation and natural fire frequency alteration; land use changes; wetlands and stream deterioration; water quality alterations; modifications of indigenous peoples' and local communities' ways of life	Generation of wildlife corridors to connect habitats; minimization of project footprint; creation of compensatory protected areas; management plans; use of native plant species as barriers to avoid or reduce undesirable intrusions; minimization of access roads and right-of-way width for pipelines; minimization of forest edges; implementation of management and maintenance plans for all routes; revegetation along all routes; right-of-way maintenance; improvement of land use management; elaboration and implementation of zoning plans; environmental education and awareness programs
Water and sanitation/flood protection	Coastal erosion downstream from river breakwaters; removal of pollutants by dredging bottom sediment; pollution of water sources; deterioration of wetlands; loss of connectivity between rivers, wetlands, and riparian zones; displacement or loss of wildlife; generation of artificial wetlands; invasions of aquatic weeds and disease vectors; worsening of water quality due to sewage disposal in water bodies; encroachment; land use changes; storm-induced floods within enclosed areas protected by dikes	Land use management; zoning; execution of pollution controls; water quality monitoring; elaboration and implementation of environmental education and awareness programs; implementation of management plans for wetland areas; maintenance of wildlife corridors, channels, and flooded areas; mechanical control of aquatic weeds; biological control of disease vectors; adequate site selection and engineering design; establishment of physical barriers; adoption of design criteria aimed at discouraging encroachment into natural habitats

Source: Quintero 2007.

CHAPTER 4

Biodiversity Conservation and Food, Water, and Livelihood Security: Emerging Issues

FOR SOME YEARS, THE WORLD BANK HAS RECOGNIZED THE THREAT that climate change poses to achieving poverty reduction and development goals. Three of the world's greatest challenges over the coming decades will be biodiversity loss, climate change, and water shortages. These three issues are closely linked to agricultural productivity and food security. The impacts on agriculture and availability of water will have the greatest potential to depress the livelihoods of the poor as well as national economic growth in the least-developed countries, especially in Africa. Recent studies show that farming, animal husbandry, informal forestry, and fisheries make up only 7.3 percent of India's gross domestic product (GDP), but these activities constitute 57 percent of GDP of the poor, who are most reliant on natural resources and ecosystem services (Sukhdev 2008). In many poor regions with chronic hunger, achieving the Millennium Development Goal of reducing poverty will require harnessing ecosystem services and rehabilitating degraded lands and natural resources critical for expanding agricultural productivity and achieving food security.

Agriculture and Biodiversity

Agriculture is one of the greatest threats to natural ecosystems worldwide. Climate change, reduced rainfall, land degradation, and rising human population pressure for lands and livelihoods are all likely to lead to agricultural expansion. Expanding agriculture will lead to further habitat loss and fragmentation, drainage of wetlands, and impacts on freshwater and marine ecosystems through sedimentation and pollution. The Millennium Ecosystem Assessment confirmed that agriculture is the dominant terrestrial influence on ecosystems and that, without major changes in current farming practices and agricultural landscape management, the agricultural frontier will likely expand and many important biological habitats will be lost. By 2050, almost 40 percent of the land currently under low-impact forms of agriculture could be converted to more intensive forms of agriculture, forcing poor farmers to open up ever more marginal lands, with further loss of biodiversity and ecosystem services (Sukhdev 2008).

Although some natural habitats have been successfully converted to productive and sustainable agricultural lands (such as the conversion of temperate forests in Europe to fertile farmland), other ecosystems have much less fertile soils and cannot support long-term agriculture. Clearance of tropical forests on low-nutrient soils, for instance, provides new land for short-term crops, but after a few years such lands lose their productivity, forcing farmers to clear more forests to open up new fields. Agricultural encroachment in such regions is likely to lead to further cycles of land degradation and abandonment.

Although agriculture is the greatest threat to biodiversity, it is also highly dependent on soil biodiversity, agrobiodiversity (crop varieties), and the ecosystem services and benefits that natural habitats provide. Collectively, agriculture benefits from the following ecosystem services:

- Regulation of water flow for downstream agriculture
- Nutrient cycling, such as decomposition of organic matter
- Nutrient sequestration and conversion, as in nitrogen-fixing bacteria
- Regulation of soil organic matter and soil water retention
- Regulation of pests and diseases
- Maintenance of soil fertility and biota
- Pollination by bees and other wildlife.

Understanding the contribution of ecosystem services to agricultural productivity and integrating protection of natural habitats into agriculture planning can contribute to sustained production even under uncertain climatic conditions (see box 4.1).

Impacts of Climate Change on Agriculture

Changing climate and rainfall patterns are expected to have significant impacts on agricultural productivity, especially in arid and semiarid regions. One study estimates that climate change could lead to a 50 percent reduction in crop yields for rain-fed agricultural crops by 2020. Most climate modeling scenarios indicate that the drylands of West and Central Asia and North Africa, for instance, will be severely

> **BOX 4.1**
> **Insects and Orange Juice:**
> **Paying for Ecosystem Services in Costa Rica**
>
> In Costa Rica, the Del Oro Company, a large producer of citrus juices, is leading the way in maintaining a balance between agriculture and nature. Its collaboration with the government of Costa Rica in conserving tropical forests in the Guanacaste National Park ensures the provision of essential ecosystem services to the plantations.
>
> The Area de Conservación Guanacaste (ACG) encompasses a range of tropical forest habitats including a belt of transition forests between the dry forests of Guanacaste and the wetter Caribbean rain forests. Approximately 1,200 hectares of the dry-wet transition forests form a wide peninsula extending into the Del Oro plantations and adjoining the ACG forests at the southern boundary of Del Oro lands. Del Oro recognizes that the ACG provides essential ecosystem services, in the form of pollination and pest control, to the citrus plantations and juice production industry. Through an agreement with the Ministry of Environment and Energy signed in August 1998, Del Oro agreed to pay for such services:
>
> - Biological control agents, primarily parasitic wasps and flies of importance to integrated pest control, were valued at $1 per hectare per year for the 1,685 hectares of Del Oro orange plantations adjacent to the AGA, for a total of $1,685 a year.
> - Water from the Upper Río Mena Basin, in the ACG, services Del Oro farms and was valued at $5 per hectare per year for the 1,169 hectares, totaling $5,885 a year.
> - Biodegradation of the orange peels from Del Oro on ACG lands was valued at $11.93 per truckload, for a minimum payment of 1,000 truckloads per year, for a total of $11,930 a year.
>
> In addition, the agreement leaves room for a possible carbon fixation program in these 1,200 hectares of wild lands and stipulates that any carbon credits will be divided equally between Del Oro and the ACG. Under the contract, the plantation agrees to maintain good agricultural practices in its plantations according to the standards and legislation of Costa Rica and the U.S. Food and Drug Administration. The agreement provides an interesting model, illustrating how recognition of ecosystem services can play a valuable role in conservation and adaptation.
>
> *Source:* Janzen 1999.

affected by droughts and high temperatures in the years to come. Droughts and flash floods have become more frequent in recent years in these regions. These largely rain-fed agricultural areas are the most vulnerable to the impact of climate change.

According to crop-climate models, in tropical countries even moderate warming can reduce yields significantly (1° C for wheat and maize and 2° C for rice) because many crops are at the limit of their heat tolerance. For temperature increases above 3° C, yield losses are expected to occur everywhere and to be particularly severe in tropical regions (World Bank 2008b). Areas most vulnerable to climate change—centered in South Asia and Sub-Saharan Africa—also have the largest number of rural poor and rural populations dependent on agriculture. Global warming, and less predictable rainfall patterns, will have a notable impact on arid and semiarid lands, many of which are already marginal for agriculture. Climate change will lead to water scarcity, increased risk of crop failure, pest infestation, overstocking, permanent degradation of grazing lands, and livestock deaths. Such impacts are already imposing severe economic and social costs and undermining food security, and they are likely to get more severe as global warming continues. This makes climate change a core development problem and ecosystem-based approaches a critical part of the solution (see box 4.2).

The Bank's response to the threats to agriculture that are presented by climate change focus on both mitigation and adaptation and can be divided into four strategic objectives:

- Monitoring impacts of climate change on crops, forests, livestock, and fisheries (adaptation)
- Providing risk management strategies for farmers and lenders against the impacts of climate change (adaptation)
- Preventing crop and livestock losses due to changing climatic factors and increased pressure from pests through improved management techniques and tolerant crop varieties and livestock breeds (adaptation)
- Improving land and resource management to maintain sustainable production (mitigation).

The Bank has a large and expanding portfolio of agriculture projects. Few projects explicitly target biodiversity conservation or ecosystem services, although many promote more sustainable agricultural practices, such as rotational cropping, reduced tillage, and soil conservation measures, which are more ecologically friendly and designed to boost yields. During the last decade, the Bank has been developing a suite of pilot conservation projects that target agriculture in, and around, protected areas or in larger landscapes of conservation interest. Such projects usually try to change production practices to provide greater biodiversity benefits (such as promotion of shade coffee) or attempt to substitute other income-earning opportunities for harmful agricultural practices. Some promote more ecosystem-friendly policies in the agriculture sector, such as integrated pest management in Indonesia to reduce dependence on high levels of pesticides.

In response to climate change, the Bank is encouraging more sustainable agriculture to avoid overgrazing and land degradation and is promoting new agroforestry systems and multispecies cropping. Increased attention is also being paid to conserving agrobiodiversity in crop gene banks and traditional agricultural practices, which maintain diversity of varieties and crops for food security (see box 4.3).

BOX 4.2
Water Tanks for Irrigation in Andhra Pradesh, India

In the Godavari River Basin in India, home to 63 million people, nearly all rain falls in the monsoon from June to October, making storage essential for year-round access to water. Poverty, limited water supplies, drought, costs of seeds and farm chemicals, and iniquitous financing by suppliers jeopardize the lives of many farmers and have resulted in a wave of farmer suicides. Climate change adds uncertainty to the frequency and rate of precipitation in the region, putting an additional burden on these farmers.

Ancient village earth dams (1–10 hectares in size), which used to function as storage tanks, have deteriorated due to mismanagement and full diversion of river water. Loss of surface waters has driven the overexploitation of groundwater, further threatening security of supply. To meet the growing demand for irrigation water, the Andhra Pradesh government proposed building a $4 billion Polavaram Dam on the Lower Godavari River, which would displace 250,000 people and inundate key habitats, including 60,000 hectares of forest.

A World Wide Fund for Nature (WWF) pilot project developed in 2004 in collaboration with a local NGO and villages assessed the costs and benefits of restoring the old water tanks. Between 2005 and 2006, in Sali Vagu subcatchment, on a tributary of the Godavari, 12 tanks with an area of 11 hectares and serving 42,000 people were restored through de-silting to capture and store more monsoon runoff. The $103,000 intervention was undertaken with funding of $28,000 in cash from WWF and $75,000 from farmers in cash inputs and labor. The increased water supply and groundwater recharge resulted in less groundwater pumping. Water tables rose, reactivating some dry wells worth an average value of $2,330 each. An additional 900 hectares were irrigated, and the nutrient-rich silt was spread over 602 hectares. Crop yields rose significantly, increasing total production by Rs 5.8 million ($69,600) per year. Irrigation of additional lands reduced the need for electricity to pump groundwater, and wages paid for de-silting the tanks supplemented farmers' incomes. In addition, use of some ponds for fish production provided a further net profit of Rs 160,000 ($3,700). The project also created artificial habitats for migratory and water birds.

The pilot project demonstrated the potential for tank restoration to meet India's soaring demand for water, in place of proposals for large-scale water infrastructure developments. In the Maner River Basin, 6,234 water tanks covering 588 square kilometers could be de-silted at an estimated cost of Rs 25.5 billion ($635 million). These could store an extra 1.9 billion cubic meters of water (compared to estimated water use in the basin today of 2 billion cubic meters per year) at a cost of $0.32 per cubic meter. Further, this water would be stored widely across the basin where more people could access it. In contrast, the government's proposed $4 billion Polavaram Dam would store 2.1 billion cubic meters of irrigation water at a cost of $1.88 per cubic meter.

Source: WWF 2008.

> **BOX 4.3**
> **Adaptation to Climate Change: Exploiting Agrobiodiversity in the Rain-fed Highlands of the Republic of Yemen**
>
> Communities in the highlands of the Republic of Yemen retain old crop varieties and traditional knowledge related to the use of these agrobiodiversity resources. Knowledge and practice have evolved over more than 2,000 years to increase agricultural productivity in areas of limited rainfall. The construction and management of terraces, for instance, help to improve the efficient use of water and to minimize land degradation. Most of the landraces and local crop varieties have been selected to meet local needs and have adaptive attributes for coping with adverse environmental and climatic conditions. The Republic of Yemen is considered an important primary and secondary center of diversity for cereals, so these crops are important genetic resources. This local agrobiodiversity is, however, threatened by global, national, and local challenges, including land degradation, climate change, globalization, anthropogenic local factors, and loss of traditional knowledge.
>
> A $4 million GEF-supported project, currently under preparation, aims to enhance coping strategies for farmers who rely on rain-fed agriculture in the Yemen highlands. The project focuses on the conservation and use of biodiversity important to agriculture (particularly local landraces and their wild relatives) and associated local traditional knowledge. This GEF project will

Sustainable Land Management

Land degradation diminishes biological diversity and many of the ecosystem goods and services on which human societies depend. Up to 75 percent of Africa's poor lives in rural areas with livelihoods critically dependent on efficient use of increasingly scarce land, water, and nutrients. Land degradation marginalizes efforts to secure long-term food security, rural productivity, and development. Climate change is likely to put further stress on fragile ecosystems. Desertification in some regions is already triggering large-scale migrations, instability, and violent conflicts over scarce resources.

As one of the leading financiers of measures to combat land degradation and desertification, the Bank continues to invest in activities that promote appropriate sustainable land management practices and protect biodiversity and ecosystems. Regional and national investments planned under the TerrAfrica umbrella are expected to improve land use practices and carbon sequestration, while promoting more sustainable land management and biodiversity. The Bank is assisting several countries in Sub-Saharan Africa, including Burundi, Ethiopia, Madagascar, Mauritania, and Senegal, in efforts to integrate sustainable land

> complement a loan through the Rain-fed Agriculture and Livestock Project. Since women do much of the farm work in the Republic of Yemen, the project will have a strong gender emphasis. The project will have four components:
>
> - *Agrobiodiversity and local knowledge assessment.* Document farmers knowledge on (adaptive) characteristics of local landraces and their wild relatives in relation to environmental parameters to develop vulnerability profiles for the crops.
> - *Climate modeling assessment.* Develop initial local predictive capacity of weather patterns, climate changes, and longer-term climate change scenarios for these rain-fed areas.
> - *Enhancement of coping mechanisms.* Identify a menu of coping mechanisms (such as in situ conservation, improved terracing with soil and water conservation practices, choice of crops, and cropping patterns) designed and piloted to increase resilience of farmers to climate variability and reduce vulnerability to climatic shifts.
> - *Enabling policies and institutional and capacity development.* Improve the capacity of key line agencies and stakeholders to collect and analyze data, improve climate predictions, and create systems of information and information flow for enhanced uptake of coping mechanisms in the agriculture sector.

management into poverty reduction strategies and investments to address land degradation. New carbon markets may also afford opportunities for the Bank to invest in land rehabilitation as well as more sustainable agricultural practices to restore productive agricultural systems and alleviate poverty. Studies have shown that ecosystem-based agriculture not only improves soil fertility and has fewer detrimental effects on the environment but also can produce similar crop yields as conventional methods (see box 4.4).

As agricultural programs take account of climate change and changing rainfall patterns, increasing emphasis is being placed on community-driven development. In Karnataka, India, farmers rely on rain-fed agriculture and a narrow range of two to five crops. Frequent droughts and poor agriculture and watershed management have led to deterioration of lands, further reducing their productivity. In 2001 the Bank funded a project in five districts to promote better management of the watershed and the associated natural resources. The project focused on soil and water conservation on 432,000 hectares of arable and non-arable land by introducing new approaches for community-based participatory planning. Project results included an increase in the availability of groundwater from four to six months and an increase in crop diversity and crop yield by 24 percent.

> **BOX 4.4**
> **Conservation Farming in Practice in South Africa**
>
> A GEF-funded medium-size project showed that conservation farming on some South African farms can reduce input costs, increase profits, and improve sustainability. These farming practices also conserve biodiversity, contribute to carbon sequestration, and improve the quantity and quality of water flow.
>
> *Farming for Flowers on the Bokkeveld Plateau*
>
> From the western rim to the eastern margin of the Bokkeveld plateau, rainfall declines from 500 millimeters to 200 millimeters per year over a distance of 15 kilometers. Over this transition, the vegetation changes from fynbos on infertile sandy soils to renosterveld to Succulent Karoo. The area supports about 1,350 plant species, 97 of which are endangered. The small village of Nieuwoudtville on the Bokkeveld plateau is the "bulb capital of the world," with a staggering 241 bulb species. The richest concentration of bulbs, both in terms of species and individuals, occurs on the highly fertile clays. Unfortunately, large areas of bulb-rich veld have been ploughed up and replaced with cereals and pasture crops.
>
> About 30 years ago, one farmer—Neil McGregor, on the farm Glen Lyon—decided that this form of agriculture was not sustainable. Instead, he began to nurture the indigenous veld to provide better plant cover. With the diversity of indigenous plants, McGregor was able to maintain productivity for much longer through the dry summer season than his neighbors did with their planted crops. By using biodiversity-friendly practices and refraining from the use of pesticides, he boosted sheep productivity and reduced the use of inputs. Moreover, he found that aardvark and porcupine, which are considered troublesome on crop farms, promoted the proliferation of bulbs and hence forage for his livestock. He abandoned attempts to control these so-called problem animals. One consequence of this conservation farming was unparalleled displays of

In Central America, the Bank has been supporting improved livestock management linked to payments for ecosystem services. The large-scale conversion of forests to pastures in Central America has resulted in the loss of biodiversity and the disruption of ecological processes. Pastures are often poorly managed and quickly become degraded, with reduced productivity. Currently, at least 30 percent of the region's pastures are considered to be degraded and are of little economic and ecological value. A Bank-funded project is exploring the relationships among silvopastoral systems, ecosystem services, and farmer livelihoods to determine how silvopastoral systems contribute to both conservation and development goals (see box 4.5). This research is providing important information on more sustainable

wildflowers with a profusion of bulb species flowering from mid-winter through to late spring. These displays draw tourists to Namaqualand, bringing additional tourist income to the farm and district. Glen Lyon has become a role model in the region, and many farmers are now using conservation farming practices. Recently, Glen Lyon was declared a national botanical garden in recognition of its biodiversity values.

Getting the Most Out of the Veld

The semiarid area of South Africa known as the Nama Karoo is characterized by highly variable rainfall from year to year. The natural veld comprises a very diverse flora of palatable shrubs and grasses, interspersed with unpalatable shrubs. This area also supports an important livestock industry, based mainly on wool and mutton production. Over the past century, the condition of ranch land over much of the Nama Karoo has deteriorated, with the proliferation of a few unpalatable species replacing more palatable plants.

One farm in Elandsfontein in the Beaufort West district instituted a grazing regime that simulated natural conditions before farming, when the veld was grazed by migrating herds of ungulates. Livestock were separated into small units and kept in one area until that area was well grazed before being moved on. The condition of the veld improved. Livestock were forced to eat both palatable and unpalatable plant species. Since the unpalatable plants were not adapted to being grazed, they lost their competitive edge, became weakened, and declined in number. The higher number of small management areas ensured a longer period between grazing, thereby enabling much of the rangeland to recover. Studies show that implementation of this system raised productivity in the district, created an ecological buffer, and increased the resilience of the veld against drought, with benefits for both biodiversity and production.

Source: Pierce and others 2002.

land management that can contribute to biodiversity conservation and carbon storage, while improving farmers' livelihoods.

Managing Invasive Alien Species

Changing land use patterns and global warming will affect the distribution of species, exacerbate other environmental stresses, and facilitate the establishment and spread of invasive alien species. Invasives are widely regarded as the second greatest threat to biodiversity after direct destruction and fragmentation of habitats. Most introductions of exotic species to new environments have been facilitated

> **BOX 4.5**
> **Payments for Environmental Services to Protect Biodiversity and Carbon in Agricultural Landscapes**
>
> Protecting biodiversity in agricultural landscapes is important both in its own right and as a means to connect protected areas, thus reducing their isolation. The challenge is finding ways to do so. The GEF-financed project Regional Integrated Silvopastoral Approaches to Ecosystem Management was implemented in Colombia, Costa Rica, and Nicaragua from 2002 to 2008 as a pilot project to demonstrate and measure the effects of paying incentives to farmers in exchange for environmental services. By the time it closed in January 2008, the project had clearly demonstrated that silvopastoral practices generate substantial benefits in biodiversity conservation, carbon sequestration, and water services and that payments can induce substantial land use changes that are environmentally beneficial.
>
> Silvopastoral production systems, which combine trees with cattle production, provide an alternative to current livestock production practices and can help to improve the sustainability of cattle production and farmer income, while providing an environment that is more hospitable to biodiversity. The project resulted in substantial carbon sequestration, both directly (by sequestering carbon in trees) and indirectly (by inducing lower applications of nitrogen fertilizers and, through improved nutrition, reducing methane emissions from livestock). Silvopastoral systems incorporate deeply rooted, perennial, native, naturalized, multipurpose, and timber tree species that are drought tolerant and retain their foliage in the dry season. As such, they provide large amounts of high-quality fodder and shade that result in stable production of milk and beef, maintain the animals' condition, and secure farmers' assets. Under extreme conditions of climate change affecting temperatures and rainy seasons, cattle ranching in pastures without trees would be more vulnerable than in pastures with trees.
>
> New projects are now being prepared in Colombia and Nicaragua to scale up and adopt biodiversity-friendly silvopastoral production systems on a larger scale. The program will help to address climate change and its consequences in the livestock sector, among other environmental and socioeconomic benefits.

by human agency, either deliberately (for example, through agricultural introductions) or accidentally (for example, in the ballast water of ships). The spread of invasives is on the rise globally, facilitated by increasing trade, tourism, international traffic, and even development assistance. Although such species may provide some immediate short-term benefits, they often entail long-term environmental and economic costs.

The threats to agricultural productivity posed by invasives (weeds, pests, and diseases of crops and livestock) have long been recognized. In recent years, the understanding of their impacts on natural ecosystems, ecosystem services, and wider human livelihoods has improved. For example, exotic plants can come to dominate freshwater bodies and waterways, affecting nutrient dynamics, oxygen availability, food webs, and fisheries. Other invasives, from microbes to mammals, pose a major threat to human health and livelihoods. Their economic impacts are large, costing an estimated $140 billion annually in the United States. Water hyacinth in Lake Victoria threatens local fisheries, and its control and removal cost around $150 million per year. Donkeys and goats cause soil degradation in parts of the Galapagos Islands, threatening fragile ecosystems, endemic species, and the local tourist economy; their removal costs more than $8 million annually (Murphy and Cheesman 2006).

The introduction of new and adaptable exotic species for agriculture to meet the growing demand for biofuels, mariculture, aquaculture, and reforestation presents a particular challenge. Ironically, the very characteristics that make a species attractive for introduction under development assistance programs (fast growing, adaptable, high reproductive output, tolerant of disturbance and a range of environmental conditions) often are the same properties that increase the likelihood of the species becoming invasive. Development programs for agriculture, especially agroforestry and aquaculture, have thus facilitated both deliberate, and unintentional, spread of invasives. Such events are costly; indeed, their negative effects may be far greater, and longer lasting, than the positive impacts of the aid programs from which they arose. Invasives accidentally introduced through development assistance programs include itch grass, a major weed in cereals in South and Central America, and a range of nematode pests. Problems resulting from intentional introductions under development assistance programs include *Tilapia* fish for aquaculture in Central America and various trees and shrubs introduced through agroforestry programs.

The impacts of invasives on land and water management and agriculture will be greatest in some of the poorest countries, including those in Africa, where land degradation and food security are major concerns. The Bank is working with the Global Invasive Species Programme (GISP) to understand the implications of invasive alien species on food production, food security, and health, including assessment of best-practice guidelines for avoiding the introduction of species known to be invasive. These capacity-building efforts have been complemented by specific projects to control, manage, and eradicate invasives in South Africa (wattles and pines), Lake Victoria (water hyacinth), India, the Seychelles, and South and Central America.

Climate change is likely to exacerbate the spread of invasives, with serious environmental and economic consequences. Invasives are already a serious problem in some vulnerable habitats such as the Cape Floristic Region in South Africa. An estimated 43 percent of the Cape Peninsula is covered in alien vegetation, consuming up to 50 percent of the region's river runoff. The availability of freshwater is a key factor limiting development in the Western Cape; where water is available, it is already fully used for agricultural, industrial, and domestic use. The spread of exotic trees in the mountain catchment areas surrounding Cape Town could reduce water resources for this rapidly growing city by another 30 percent. These losses could mean that more (and expensive) dams have to be built to meet the demand for water. Economic studies have shown that clearing invasive species in the catchment areas will increase water production and deliver water supplies at much less cost than building a new reservoir (see box 4.6).

Additionally, invasive plants in indigenous grasslands and scrublands increase fuel loads and fire risk, which leads to increased soil erosion, degradation, and loss of biodiversity and ecosystem services in mountain catchments. The South African government has taken serious action to address these threats through the Working for Water and the Working for Fire programs, which are collaborating with the Bank–Global Environment Facility (GEF) CAPE Biodiversity Conservation and Sustainable Development Project to better manage and control invasive species in the Cape Floristic Region. Working for Water brings additional benefits through increased employment opportunities for disenfranchised groups. Support to Working for Water from the Bank's Development Marketplace has increased employment opportunities for marginalized people through small-scale industries that use the harvested alien trees.

BOX 4.6
A Cost-Effective Solution for Increasing Water Supply: Removing Invasive Species in South Africa

South Africa has a serious problem of invasive alien plants that affects 10 million hectares (8.28 percent) of the country. These invasions come at a significant ecological and economic cost. Invasive species, with their high evapotranspiration rates, are an immense burden to already water-scarce regions. Numerous studies have analyzed the role of invasive alien plants in decreasing the amount of water available to reservoirs. In 2002 the South African government approved R 1.4 billion ($173.5 million) for the proposed Skuifraam Dam Project on the Berg River near Franschhoek to help address the looming water crisis in the Western Cape and Cape Town. A feasibility study for the planned dam demonstrated that water delivery would cost 3 cents less per kiloliter if invasive species were managed in the catchment area. It was estimated that clearing invasive plants from the Theewaterskloof catchment would deliver additional water at only 10.5 percent of the cost of delivery from the new Skuifraam scheme if no clearance was carried out. Accordingly, large-scale programs to clear invasive trees are being undertaken as part of management for the new Berg Dam.

Source: Pierce and others 2002.

Protecting Natural Ecosystems for Water Services

Water is essential for all life on Earth. The impacts of climate change can be expected to have serious consequences on the availability and quality of water resources. Melting glaciers, higher intensity and more variable rainfall events, and rising temperatures will contribute to increased inland flooding, increased water scarcity, and decreasing water quality. Restoration and maintenance of watersheds, including management of soils, can contribute to reducing the risk of flooding and maintaining regular water supplies. Natural ecosystems such as wetlands and forests act as natural water recharge areas, storing runoff, recharging aquifers, and replenishing stream flows. This reduces the risk of floods associated with heavy rainfall or glacier melt. A study of upland forests in a watershed in Madagascar has estimated the annual value of flood protection afforded by these forests at $126,700 (Kramer and others 1997; see box 4.7).

BOX 4.7
The Downstream Benefits of Forest Conservation in Madagascar

Economic analysis can be a useful tool for demonstrating the social benefits of protected areas and conservation. A World Bank study showed that the economic benefits of biodiversity conservation far outweigh the costs in Madagascar. Sustainable management of a network of 2.2 million hectares of forests and protected areas over a 15-year period was estimated to cost $97 million (including opportunity costs forgone in future agricultural production) but to generate $150 million to $180 million in total benefits. About 10-15 percent of these benefits are from direct payments for biodiversity conservation, 35-40 percent from ecotourism revenues, and 50 percent from watershed protection, primarily as a result of maintaining water flows and averting the impacts of soil erosion on smallholder irrigated rice production.

The study considered potential winners and losers from forest conservation, pointed to the need for equitable transfer mechanisms to close this gap, and emphasized the role of conservation in helping to maintain or improve the welfare of at least half a million poor peasants. The study contributed to a government decision to expand forest protected areas to more than 6 million hectares in Madagascar. The Bank and other donors are helping to fund the expanded protected area network through the Third Environment Program, including capitalization of a conservation trust fund to provide sustainable financing. Carbon finance will also provide support to protect Madagascar's rich forests and the unique lemurs and other endemic fauna for which the island is famed.

Similarly, Sri Lanka's Muthurajawela marsh, a coastal peat bog covering some 3,100 hectares, is an important part of local flood control. The marsh buffers floodwaters from the Dandugam Oya, Kala Oya, and Kelani Ganga rivers and discharges them slowly into the sea. The annual value of these services has been estimated at more than $5 million, or $1,750 per hectare of wetland area (Emerton and Bos 2004). Natural wetlands are also part of water treatment and flood control strategies in the Yangtze Basin in Hubei Province (see box 4.8).

Rising temperatures and the growing need for irrigated agriculture will increase the pressure on scarce water resources. Overall, the greatest human requirement for freshwater resources is for crop irrigation, particularly for farming in arid regions and in the great paddy fields of Asia. In Asia, irrigated lowland agriculture in the large basins receiving runoff from the Hindu Kush–Himalayan system is projected to suffer from lack of water in the dry season. In South Asia, hundreds of millions of people depend on perennial rivers such as the Indus, Ganges, and Brahmaputra, all fed by the unique water reservoir formed by the 16,000 Himalayan glaciers. The current trends in glacial melt suggest that low flows will be reduced

---

BOX 4.8
Lakes in the Central Yangtze River Basin, China

In 2000 China's central government ordered all cities of more than 500,000 people to treat at least 60 percent of their wastewater. As part of that order, the government endorsed a $4.5 billion scheme to build 150 new wastewater treatment plants along the Yangtze River by 2009.

A pilot plant for this project is located in Chongqing, in Sichuan Province. Chongqing lies in the basin of the Yangtze River and is the largest municipality in China, generating nearly 1 billion tons of untreated wastewater a year. The pilot plant has been operational for a year and provides primary treatment to more than 50,000 cubic meters of water a day. This treatment involves multiple screens that remove large debris and an ultraviolet disinfection mechanism that reduces microorganisms. Due to relatively high installation costs, the treatment plant does not include systems to remove organic pollutants or dissolved nitrogen and phosphorus, which increases the risk of nutrient pollution in the surrounding waters.

In the same river basin, water quality in the neighboring province, Hubei, has been deteriorating over the last 50 years. Within Hubei, however, natural ecosystems have been integrated into water treatment strategies. Previously, 757 out of 1,066 lakes had been converted to polders, reducing wetlands area 80 percent and flood retention capacity 75 percent. Application of fertilizers to aquaculture pens contributed to pollution of the lakes. The loss of connection to the Yangtze River prevented diluting flows and migration of fish. Damage

even further as a consequence of climate change. In addition, an increase in agricultural demand for water by 6 to 10 percent or more is projected for every 1° C rise in temperature. As a result, and even under the most conservative climate projections, the net cereal production in South Asian countries is projected to decline between 4 and 10 percent by the end of this century (IPCC 2007).

Retreating glaciers are also a serious concern in the Andes. As a part of the Adaptation to the Impact of Rapid Glacier Retreat in the Tropical Andes Project, which started in May 2008, the Bank is implementing a water management plan in Peru that includes water storage infrastructure and improved water use practices in the agricultural and livestock sectors. In Bolivia, the project is incorporating the impact of rapid glacier retreat into integrated watershed management, devising an integrated pilot catchment management plan for watersheds, and mainstreaming adaptive river defenses for Huayhuasi and El Palomar settlements. The project includes specific adaptation measures such as improved streamside conservation and management and improved management of glacier buffer zones, adopting an ecosystem-based approach to adaptation. In Peru and Ecuador,

---

from four major floods between 1991 and 1998 resulted in thousands of deaths and billions of dollars in damages.

To ameliorate these conditions, government agencies and NGOs have been restoring the wetlands in the basin, reconnecting the flows between the lakes and the Yangtze River. Since 2005, the sluice gates at lakes Zhangdu, Hong, and Tien'e zhou have been seasonally reopened, and illegal and uneconomic aquaculture facilities and other infrastructure have been removed or modified. Now these 448 square kilometers of wetlands can store up to 285 million cubic meters of floodwaters, reducing vulnerability to flooding in the central Yangtze region. Cessation of unsustainable aquaculture, better agricultural practices, and reconnection to the Yangtze River have reduced pollution levels in these lakes. Pollution fell at Lake Hong from national pollution level IV (fit for agricultural use only) to level II (drinkable) on China's five-point scale.

Healthy wetlands naturally remove organic and inorganic pollutants and supply clean water. Restoration of these wetlands provided more services than constructing wastewater treatment plants and at a considerably cheaper cost. Rehabilitation of these wetlands has also considerably enhanced the biodiversity of the lakes, bringing back 12 migratory fish species. Hong Lake supported only 100 herons and egrets when polluted; after restoration, the lake supported 45,000 wintering water birds and 20,000 breeding birds, including the endangered Oriental White Stork. Similar positive results were seen in Tian'e zhou and Zhangdu lakes as well.

Source: WWF 2008.

adaptation measures include forest protection, reforestation, and forest regeneration activities, aimed mostly at conserving natural ecosystems and increasing the resilience of forest ecosystems to the impacts of climate change. By restoring and harnessing ecosystem services, the project will decrease the risks of sudden floods due to glacier melt, provide alternative water storage options, and reduce erosion and siltation.

Natural Water Towers

Growing concern over water scarcity provides a powerful argument for protection of natural habitats and creation of protected areas. Ecosystem-based approaches can form an integral part of strategies to maintain water supplies for agriculture and domestic use. Municipal water accounts for less than a tenth of human water use, but clean drinking water is a critical need. Today, half of the world's population lives in towns and cities, and one-third of this urban population lacks clean drinking water. These billion have-nots are unevenly distributed: 700 million city dwellers in Asia, 150 million in Africa, and 120 million in Latin America and the Caribbean. With expanding urban needs, cities face immediate problems related to access to clean water and mounting problems related to supply.

Among the world's largest cities, 33 out of 105 obtain a significant proportion of their drinking water directly from protected areas (Dudley and Stolton 2003). These cities include Jakarta, Mumbai (formerly Bombay), Karachi, Tokyo, Singapore, Mexico City, New York, Bogotá, Rio de Janeiro, Los Angeles, Cali, Brasilia, Vienna, Barcelona, Nairobi, Dar es Salaam, Johannesburg, Sydney, Melbourne, and Brisbane. Elsewhere, half of Puerto Rico's drinking water comes from the last sizable area of tropical forest on the island, which is in the Puerto Rico National Park. Quito, the capital of Ecuador, draws its water from a system of protected areas. Mount Kenya, the second highest mountain in Africa, is one of Kenya's five main "water towers" and provides water to more than 2 million people.

In recent years, governments and city councils began to take an increasing interest in the opportunities for offsetting or reducing some of the costs of maintaining urban water supplies—and, perhaps even more important, water quality—through management of natural resources, particularly forests and wetlands. The government of Spain is promoting reforestation of the Pyrenees to improve the quality of downstream water resources. Similarly, the values of watershed protection functions in the Philippines have been estimated at $223–$455 per hectare per year (Paris and Ruzicka 1991). In Riverside, California, the local authorities have rehabilitated a natural wetland in lieu of building a denitrification facility with considerable cost savings (see box 4.9).

Many mountain protected areas can be justified through their provision of ecosystem services, such as clean water, soil conservation, and protection of

> **BOX 4.9**
> ## Wastewater Treatment with Wetlands
> A regulatory revision required the city of Riverside, California, to remove nitrogen from its wastewater. The cost of a conventional denitrification facility at the treatment plant was estimated at $20 million. After investigating alternatives, the city decided to employ a wetland system for removing nitrogen. Hidden Valley, a low-grade wetland infested with invasive alien vegetation near the treatment plant, was cleared of invasives and rehabilitated to provide the purification treatment along with other ecosystem benefits. The cost of constructing the 28 hectare wetlands project was only $2 million, a savings of $18 million, 90 percent less than a conventional facility. The costs of operating and maintaining the wetland system are also more than 90 percent less than those of a conventional system. In operation since May 1995, the system has proven effective at removing nitrogen and has met all permit requirements. Furthermore, the wetland provides important ancillary benefits that could not be provided by a conventional facility. The wetland includes an interpretive center for environmental education and recreational trails that attract more than 10,000 visitors a year. It also supports wildlife habitat that is home to 94 bird species.
>
> *Source:* Barrett 1999.

downstream, and vulnerable, communities from natural hazards such as floods and unstable hillsides. Various Bank projects have provided funding to protected areas in forest watersheds, which safeguard the drinking supplies for some of the world's major cities. Panda reserves in the Qinling Mountains, China, protect the drinking water supplies for Xi'an. The Gunung Gede-Pangrango in Indonesia safeguards the drinking water supplies of Bogor, Jakarta, and Sukabumi and generates water with an estimated value of $1.5 billion annually for agriculture and domestic use. Similarly, Kerinci-Seblat National Park in Sumatra safeguards water supplies for more than 3.5 million people and 7 million hectares of agricultural land, while two of the Andean protected areas in Ecuador provide drinking water for 80 percent of Quito's population. The La Visite and Pic Macaya national parks in Haiti safeguard water supplies for the cities of Port au Prince and Les Cayes, respectively. In Mexico, the Monarch Butterfly Reserve protects an amazing biological phenomenon and the drinking water of Mexico City. The Aberdare Mountains and Mount Kenya national parks in Kenya provide critical water to Nairobi, while the Udzungwas in the Eastern Arc Mountains of Tanzania supply water to Dar es Salaam. Similarly, a recent study in Mongolia demonstrated that maintaining natural ecosystems in the Ulaanbaatar watershed to protect the city's water supplies makes more economic sense than allowing urban development to expand into the former reserve (see box 4.10).

> **BOX 4.10**
> **Protected Areas as Water Towers:**
> **Mongolia's Least Costly Solution**
>
> The wells that supply Ulaanbaatar with drinking and industrial water have almost reached their limit. Demand for water is fast outstripping supply. Seasonal water shortages are growing ever more common, and at some time within the next 10 years the city will face a critical shortfall in water availability. Ulaanbaatar derives its water from the Tuul Basin, which has a catchment area of almost 50,000 square kilometers, through which the river runs for a length of more than 700 kilometers. The Tuul River, its main tributary the Terelj, and another 40 smaller rivers, streams, and lakes are fed by rainfall, snowmelt, and groundwater and drain the southern slopes of the Baga Khentii to the northeast of the city.
>
> Ecological conditions in the upper watershed have a direct link to the availability of surface water and groundwater downstream in Ulaanbaatar. Natural vegetation cover is particularly critical, as it influences rainfall interception, runoff, and water discharge over the course of the year. The extent and quality of forests, grasslands, and soil cover affect the Tuul River's mean flow and flow duration, influence the timing and intensity of peak and low flows, and determine the extent and rate of groundwater recharge. They also affect the silt and sediment loads that are carried downstream. Basically, a healthy upstream ecosystem helps to ensure clean, regular, and adequate river flow and groundwater resources for Ulaanbaatar.
>
> A recent study showed that, as the ecosystem is degraded and land cover is lost, average runoff will increase, and the river's mean annual maximum and low flows will be intensified. Diminished discharge will lower the groundwater table from between 0.24 meter (under a continuation of the status quo) and 0.4 meter (under a scenario of rapid degradation). In 25 years' time, daily water supply in Ulaanbaatar will be reduced by some 32,000 and 52,000 cubic meters, respectively. In contrast, conservation and sustainable use of the upper watershed will protect current river flow and groundwater levels. Considering the gains (sustained water supplies to Ulaanbaatar) and losses (reduced land values in the upper watershed), conservation of natural habitats in the Upper Tuul is the most economically beneficial future management scenario. The conservation and sustainable use scenario yields a net present value, over 25 years, of $560 million. This is higher than the net present value generated under either a continuation of the status quo or a scenario of rapid ecosystem degradation.
>
> *Source:* Emerton and others 2009.

CHAPTER 5

Implementing Ecosystem-Based Approaches to Climate Change

CLIMATE CHANGE HAS BECOME THE KEY ENVIRONMENTAL CONCERN of the decade. Much attention is rightly focused on reducing carbon emissions and greenhouse gases (GHGs) from transport and energy sectors by reducing the use of fuel and adopting improved technologies. Nevertheless, as countries search for medium- and longer-term mitigation and adaptation measures, protection of natural habitats must be a key part of climate change strategies. The world's poorest people, who depend directly on the services that various ecosystems provide, are also the most vulnerable to the effects of climate change. This makes conservation of biodiversity, and the services that healthy ecosystems provide, a triple-A investment. Healthy ecosystems can reduce vulnerability to climate shocks, protect the web of life on which people depend for goods and services, and increase local and national resilience to the impacts of climate change.

The Bank has access to several instruments and financing mechanisms that can assist client countries to incorporate ecosystem-based solutions into climate mitigation and adaptation strategies. These include Bank programs and projects, development policy lending, the Strategic Framework for Climate Change and Development (SFCCD) and the new Environment Strategy (under preparation), as well as assistance to countries for economic sector work, strategic environmental assessments, poverty reduction strategy papers, and national adaptation strategies. In addition to Bank lending and Global Environment Facility (GEF) grants, the Bank is facilitating the development of market-based financing mechanisms and piloting new avenues to deepen the reach of the carbon market.

Looking Forward: The Strategic Framework for Climate Change and Development

The Bank recognizes that global efforts to overcome poverty and advance sustainable development must address climate change and its economic, environmental, and social implications. In order to address these questions efficiently, the SFCCD seeks to examine climate change from a multisectoral and multifaceted perspective across the institution. The SFCCD consists of the following six pillars for action:
- Support climate actions in country-led development processes
- Mobilize additional concessional and innovative finance
- Facilitate the development of market-based financing mechanisms
- Leverage private sector resources
- Support accelerated development and deployment of new technologies
- Step up policy research, knowledge, and capacity building.

In addition to focusing on immediate actions to promote cleaner and renewable energy, the SFCCD recognizes that ecosystems and biodiversity provide essential services that underpin every aspect of human life, including food security, carbon storage, climate regulation, livelihoods, ethnic diversity, and cultural and spiritual enrichment. Enhanced protection and management of natural habitats and biological resources can help to mitigate climate change; they also provide *effective* and *low-cost* options to reduce vulnerability and adapt to climate change.

Bank projects and programs are already supporting biodiversity conservation and protecting natural habitats and ecosystem services, thereby contributing to effective mitigation and adaptation strategies. Nevertheless, the Bank could, and should, support a stronger focus on ecosystem management as part of an explicit response to climate change, including the following:
- Protecting terrestrial, freshwater, and marine ecosystems and ecological corridors to conserve terrestrial and aquatic biodiversity and ecosystem services
- Integrating protection of natural habitats into strategies to reduce vulnerability and disaster risks (including protection from natural hazards such as floods, cyclones, and other natural disasters)
- Scaling up country dialogue and sector work on valuation of ecosystem services and the role of natural ecosystems, biodiversity, and ecosystem services in underpinning economic development
- Emphasizing the linkages between protection of natural habitats and regulation of water flows and water quality for agriculture, food security, and domestic and industrial supplies

- Scaling up investments for protected areas and natural ecosystems linked to sector lending, such as infrastructure, agriculture, tourism, water supply, fisheries, and forestry
- Promoting greater action on management of invasive alien species, which are linked to land degradation and have a negative impact on food security and water supplies
- Emphasizing the multiple benefits of forest conservation and sustainable forest management (carbon sequestration, water quality, reduction of the risks from natural hazards, poverty alleviation, and biodiversity conservation)
- Promoting investments in natural ecosystems as a response to mitigation (avoided deforestation) and adaptation (wetland services)
- Integrating indigenous crops and traditional knowledge on agro-biodiversity and water management into agricultural projects as part of adaptation strategies
- Promoting more sustainable natural resource management strategies linked to agriculture, land use, habitat restoration, forest management, and fisheries
- Developing new financing mechanisms and integrating ecosystem benefits into new adaptation and transformation funds
- Using strategic environment assessments as tools to promote protection of biodiversity and ecosystem services
- Monitoring investments in ecosystem protection within mainstream lending projects and documenting good practices for dissemination and replication
- Developing new tools to measure the benefits of integrated approaches to climate change (ecosystem services, biodiversity conservation, carbon sequestration, livelihood co-benefits, and resilience).

Growing Forest Partnerships

In collaboration with the Food and Agriculture Organization (FAO) and the International Union for Conservation of Nature (IUCN) and with technical support from the International Institute for Environment and Development, the Bank is supporting implementation of the Growing Forest Partnerships (GFP) initiative, which was informed by an independent, global consultation of more than 600 forest stakeholders, including a special survey of indigenous peoples. The GFP aims to facilitate bottom-up, multiple-stakeholder partnership processes in developing countries to identify national priorities and improve access to the financing available through a wide variety of international means and mechanisms, for example, carbon finance, private sector investments, and overseas development assistance. The GFP aims to provide a platform to ensure that marginalized, forest-dependent groups can participate in the formulation of national priorities and be included in the international dialogue on forests. The GFP will work through locally based institutions and will build on existing partnership structures. The Bank is supporting this initiative with start-up funding from the Development Grant Facility.

The GFP will provide a platform to achieve progress in the following target areas by 2015: (a) creating an enabling environment for carbon-based forestry activities; (b) promoting the use of forests for poverty alleviation under conditions of climate change; (c) achieving significant growth in sustainably managed, and legally traded, forest products and expanding the area of responsibly managed forests; (d) increasing the establishment, management, and financial sustainability of protected forest areas; and (e) reducing the area of primary forest converted to alternative land uses. The GFP will facilitate and scale up activities associated with implementation of the Bank's Forest Strategy. It will link existing and new partnership programs that promote enabling conditions in the forest sector (for example, the Forest Law Enforcement and Governance Initiative and the Multi-Donor Program on Forests) with the Bank's existing lending and financial instruments as well as new sources of concessional financing.

Developing Financing Mechanisms to Support Ecosystem-Based Approaches

There is growing consensus between the parties to the international conventions on climate change (the United Nations Framework Convention on Climate Change, UNFCCC) and biological diversity (Convention on Biological Diversity, CBD) on the need to strengthen conservation and management of natural ecosystems as part of climate change response strategies. The Ad Hoc Technical Expert Group (AHTEG) on Biodiversity and Climate Change was established to provide relevant information to the CBD and the UNFCCC through the provision of scientific and technical advice and assessment on integrating the conservation and sustainable use of biodiversity into climate change mitigation and adaptation activities. The AHTEG has emphasized the key roles that natural ecosystems can play in mitigation and adaptation to climate change and in protection of ecosystem services. Nevertheless, a key challenge remains: how to reward countries that conserve these natural ecosystems and provide global services.

Currently very few markets exist to provide financial benefits for improved management of natural ecosystems in the context of climate change, and most opportunities have come about through the voluntary carbon markets. The Clean Development Mechanism (CDM) under the Kyoto Protocol, for instance, gives carbon credits for forestation and reforestation projects (including natural forest regeneration) but makes no provision for protecting standing forest and other intact natural habitats. The Bank has been a leader in promoting innovative financing mechanisms to protect natural ecosystems for their carbon sequestration and biodiversity benefits. Initiatives such as the BioCarbon Fund and the Forest Carbon Partnership Facility afford opportunities to protect forests for carbon sequestration and other multiple benefits, including conservation of biologically rich habitats, and to realize greater community benefits from forest management and watershed protection. New opportunities also exist through the GEF Adaptation Fund and links to new Bank programs such as the Global Facility for Disaster Reduction and Recovery.

Under the BioCarbon Fund, the Bank is working through existing carbon markets to bring new revenue streams to rural communities through reforestation, currently the only land use or forestry activity allowed under the CDM. Through the BioCarbon Fund, the Bank has committed to purchase emissions reductions from 17 reforestation projects in developing countries, all of them expected to generate biodiversity benefits. The BioCarbon Fund is also pioneering carbon credits for soil and agriculture carbon. This activity is not allowed under the CDM at present, but it is being discussed by the UNFCCC. This would further the penetration of carbon markets into rural communities. At the same time, the BioCarbon Fund is developing methodologies to allow a robust system of carbon payments and is piloting activities in Kenya.

Climate Investment Funds

Recognizing that a future financial architecture still has to be developed and agreed upon for climate change interventions after 2012, the World Bank, jointly with the Regional Development Banks and in consultation with developed and developing countries and other stakeholders, has developed the Climate Investment Funds (CIFs), which received the formal approval of the World Bank Board of Executive Directors in July 2008. These are an interim measure to scale up assistance to developing countries for efforts to address climate change and to strengthen the knowledge base in the development community.

The CIF umbrella covers two funds: the Clean Technology Fund and the Strategic Climate Fund (SCF). Two of the pilot programs under the SCF are the Forest Investment Program (FIP) and the Pilot Program for Climate Resilience (PPCR). The PPCR will be implemented in eight vulnerable countries. It will demonstrate ways of integrating climate risk and resilience into core development planning. The PPCR will be country-led and will support country-specific plans and investment programs to address climate risks and vulnerabilities. For most of these countries, improved management of ecosystems and natural resources are important components of building resilience and reducing vulnerability in targeted sectors (see table 5.1).

Reducing Emissions from Deforestation and Degradation

Forestry, land use change, and agriculture are major issues for climate change, accounting for almost 45 percent of emissions in developing countries. Reducing emissions from deforestation and forest degradation (REDD) has been identified

TABLE 5.1
Potential Benefits from Ecosystem Protection

Country	Food Security[a]	Infrastructure[b]	Carbon Sequestration[c]	Water Security[d]	Coastal Zone Management[e]
Bangladesh	✓	✓	✓		✓
Bolivia	✓			✓	
Cambodia	✓	✓	✓		✓
Mozambique	✓			✓	✓
Nepal		✓	✓	✓	
Niger	✓				
Tajikistan	✓			✓	
Zambia	✓		✓		

a. Ecosystem-based approaches that implement crop rotations, choose crops with less intensive nutrient and water requirements, control invasive alien species, maintain local landraces and crop varieties, and protect reefs and mangroves for sustainable fisheries.
b. Planning that protects natural habitats and ecological connectivity, incorporates protection of natural ecosystems into coastal defenses and flood control (rather than relying solely on infrastructure such as seawalls and drainage canals), and accommodates ecological flows and ecosystem functions in reservoir and dam design.
c. Reduction of carbon emissions through ecosystem-based approaches, such as establishment of new protected areas and improved management of existing reserves; protection of old-growth and swamp forests and wetlands; and natural regeneration of forests, reforestation, and afforestation.
d. Ecosystem-based approaches that include watershed and forest protection, incorporate wetlands in water treatment and water quality improvement initiatives, and protect wetlands for water storage and flood control.
e. Management that incorporates mangroves and other coastal wetlands into storm protection and coastal defense; protects mangroves, sea grass beds, and coral reefs for sustainable fisheries; and promotes integrated coastal management to prevent pollution of the marine and coastal environment.

as one of the most cost-effective ways to lower emissions (Stern 2007). Under the CDM of the Kyoto Protocol, countries cannot receive credits for REDD. However, REDD holds promise for linking carbon to improved biodiversity conservation and related benefits, since it relies on protection and improved management of natural forests.

There is some controversy over how REDD should be funded and how emissions will be measured and monitored. Ascertaining deforestation trends is difficult, especially if payments are linked to incremental reductions in rates of deforestation. The IPCC has provided guidelines for monitoring and measuring GHG emissions from deforestation and forest degradation, and more recently the World Bank and the United Nations Environment Programme have presented a concept paper to GEF on developing standard measures and models for carbon sequestration and storage. A trading mechanism would allow developing countries to sell carbon credits on the basis of successful reductions in emissions from deforestation; such credits would probably relate to national emissions and not be linked to individual sites. Any such mechanisms could generate significant additional funding for forest protection, perhaps as much as $1.2 billion a year. This is considerably more than the estimated $695 million spent annually on all protected areas (not only in forest ecosystems) in developing countries. In contrast, forestry exports from the developing world were worth more than $3.9 billion in 2006. REDD could provide strong incentives for forest conservation but is unlikely to benefit all forests equally. For REDD to contribute to combating climate change, countries would need to target threatened forests with a high volume of carbon in their biomass and soils. High-priority sites for tackling deforestation to reduce emissions may not always reflect other forest values such as biodiversity conservation, livelihood benefits, or water delivery (Miles and Kapos 2008).

One obvious risk associated with REDD is the displacement of pressures resulting from continuing demand for land for agriculture, timber, and even biofuels to ecosystems with low carbon values, either less carbon-rich forests or non-forest ecosystems such as savananas or wetlands. Another key issue is the question of who owns carbon and who should benefit from any carbon credits: national governments or the local communities and indigenous groups who manage and protect those forests and are dependent on them for their livelihoods. Assuring the equitable distribution of revenues gained from carbon credits to communities affected by improved forest protection may prove to be a key challenge of REDD implementation (see box 5.1). Implementing REDD successfully will require agreement on clear goals, eligibility criteria, and priorities as well as strong national and international capacity to monitor, manage, and evaluate performance over time.

Forest Funds

Recognizing the importance of the REDD mitigation strategy, the World Bank established the Forest Carbon Partnership Facility (FCPF) to build the capacity of developing countries in the tropics to tap into financial incentives for REDD under future regulatory or voluntary climate change regimes. The FCPF has dual objectives: to build capacity for REDD in developing countries and to test performance-based incentive payments on a relatively small scale in some pilot countries. The FCPF became operational in June 2008 with the start of operations of the Readiness Mechanism, which was triggered by capitalization of the Readiness Fund at the required minimum ($20 million); today donors have contributed $55 million to the Readiness Fund and $21 million to the Carbon Fund. The Readiness Fund will finance activities designed to (a) establish a national reference scenario for emissions, (b) adopt national REDD strategies, and (c) design national monitoring systems.

> **BOX 5.1**
> **Principles for Leveraging Benefits from REDD for the Poor**
>
> 1. *Provide information.* Basic details of how REDD mechanisms work, realistic expectations of benefits, and possible implications of different approaches are required.
> 2. *Provide up-front finance and mechanisms to reduce costs.* Provision of up-front finance would significantly improve the equitable distribution of benefits; for example, at community levels, some options for self-financing could be explored, such as improved agricultural production, nonfarm employment, and revolving credit programs.
> 3. *Use "soft" enforcement and risk reduction measures.* "Hard" enforcement measures such as financial penalties are likely to affect the poor disproportionately. Instead, "soft" measures such as nonbinding commitments to emissions reduction should be applied where possible.
> 4. *Prioritize "pro-poor" REDD policies and measures and long time horizons.* Stable and predictable benefits would provide increased security to the poor.
> 5. *Provide technical and legal assistance.* To ensure "voice and choice," improved access to appropriate legal support is crucial for poor people.
> 6. *Maintain flexibility in the design of REDD mechanisms.* Flexibility is crucial in order to minimize risks such as communities being locked into inappropriate long-term commitments.
> 7. *Clearly define and equitably allocate carbon rights.* Rights to own and transfer carbon are essential, and such rights are likely to govern land management over long time scales.
> 8. *Develop social standards.* Social standards would improve benefits for the poor by ensuring that processes are transparent. Standards should also be developed for ongoing social impact assessments.
> 9. *Apply measures to improve the equity of benefit distribution.* Issues such as baseline setting, risk aversion, and cost-effectiveness can lead to variable distribution of benefits.
> 10. *Align REDD strategies with international and national financial and development strategies.* Aligning REDD schemes with existing development processes such as poverty reduction strategies would help to raise the profile of the poor.
>
> *Source:* Peskett and others 2008.

Initially, 25 countries were accepted into the facility: Cameroon, the Democratic Republic of Congo, Ethiopia, Gabon, Ghana, Kenya, Liberia, Madagascar, the Republic of Congo, and Uganda in Africa; Argentina, Bolivia,

Colombia, Costa Rica, Guyana, Mexico, Nicaragua, Panama, Paraguay, and Peru in Latin America and the Caribbean; and Lao People's Democratic Republic, Nepal, Papua New Guinea, Vanuatu, and Vietnam in Asia. As of March 2009, 12 more countries have been added to the list, bringing the number of participating countries to 37.

Within the framework of the Strategic Climate Fund, targeted programs can be established to provide financing to pilot new development approaches or scaled-up activities aimed at a specific climate change challenge or sectoral response. The Forest Investment Program (FIP), proposed at the June 2008 CIF design meeting in Potsdam, Germany, is a program under the SCF that will mobilize significantly higher investments to reduce deforestation and forest degradation and promote improved sustainable forest management, leading to emission reductions and the protection of carbon reservoirs. The FIP will take into account country-led strategies for the containment of deforestation and degradation and build on complementarities between existing forest initiatives.

Apart from carbon and climate funds administered through the World Bank Group, the Bank is collaborating with the Congo Basin Forest Fund led by the African Development Bank to build national and local capacity for sustainable forest management in the Congo Basin and with the Norwegian International Climate and Forest Initiative, launched in December 2007, to reduce GHGs from deforestation of tropical forests in developing countries. The Bank is also represented on the steering committee of the United Nations Collaborative Program on Reducing Emissions from Deforestation and Forest Degradation in Developing Countries (UN-REDD). The first phase of UN-REDD, with financial contributions from Norway, will help to develop national strategies, establish systems for monitoring forest cover and biomass, and report on emission levels and general administrative capacity building in select countries (Bolivia, the Democratic Republic of Congo, Indonesia, Panama, Papua New Guinea, Paraguay, Tanzania, Vietnam, and Zambia). The Bank is also collaborating with the Prince of Wales Rainforest Trust on proposed REDD initiatives and the creation of green bonds to fund future investments in tropical forest conservation.

This wide range of forest initiatives and new financing mechanisms provides exciting opportunities for improving the conservation and management of natural ecosystems, especially tropical forests, with associated benefits expected for many species, habitats, and ecosystem services. Nevertheless, it is unlikely that any international mechanism linked to the UNFCCC will explicitly support forest ecosystem services other than carbon storage (see box 5.2). Under such circumstances, it may be more efficient to focus limited conservation funds on ecosystems other than forests or on forests with low carbon content rather than on high-biodiversity forests that could be covered by REDD mechanisms.

> **BOX 5.2**
> ## Can Carbon Markets Save Sumatran Tigers and Elephants?
>
> Riau Province in central Sumatra harbors populations of the critically endangered Sumatran tiger and the endangered Sumatran elephant within a high-priority Tiger Conservation Landscape. Riau has lost 65 percent of its original forest cover and has one of the highest rates of deforestation in the world, due to loss and conversion of forest for agriculture, for pulpwood plantations, and for expanding industrial oil palm plantations to serve the surging biofuels market. If the current rate of deforestation continues, estimates suggest that Riau's natural forests will decline from 27 percent today to only 6 percent by 2015. All of this comes at a global cost. The average annual carbon dioxide (CO_2) emissions from deforestation in Riau exceed the emissions of the Netherlands by 122 percent and are about 58 percent of Australia's annual emissions. Between 1990 and 2007, Riau alone produced the equivalent of 24 percent of the targeted reduction in collective annual GHG emissions set by the Kyoto Protocol Annex I countries for the first commitment period of 2008–12.
>
> Can carbon trading provide a new economic incentive to protect Riau's forests, especially the carbon-rich peat swamp forests? At present, countries are not rewarded for retaining forest canopy (avoided deforestation); instead, the emphasis is on afforestation. Second, although new programs are under consideration to provide incentives for conserving forests, the prevailing price of carbon may be too low to shift incentives from forest clearance for biofuels or pulp to conservation. Third, even if the price of carbon rises sufficiently, Riau's forests may not be given priority over other forests with higher carbon sequestration potential because the proposed new systems pay only for carbon, giving little attention to the biodiversity value of forests.
>
> Yet carbon markets may have potential to promote conservation in less productive lands. In parts of South Asia, the returns (present value) of arable land are often as low as $100 to $150 per hectare. Clearing a hectare of tropical forest could release 500 tons of CO_2. At an extraordinarily low carbon price of even $10 per ton of CO_2, an asset worth $5,000 per hectare is being destroyed for a less valuable use. A modest payment through avoided deforestation schemes could be sufficient to shift incentives in some of the unproductive arable land in South Asia.
>
> *Source:* Damania and others 2008.

APPENDIX

Securing Carbon Finance at the World Bank: Minimum Project Requirements

Type of Project

- Greenhouse gases targeted should be those covered under the Kyoto Protocol (carbon dioxide, nitrous oxide, methane, hydrofluorocarbons, perfluorocarbons, and sulfur hexafluoride).
- The Carbon Finance Unit, in accordance with the Marrakesh Accords, can support afforestation and reforestation projects in non-Annex I countries and a whole range of land use, land use change, and forestry projects in Annex I countries.

Adequate Volume of Emission Reductions

- The volume of emission reductions must be large enough to make a project viable under the Clean Development Mechanism (CDM)—for example, a small-scale project should generate a minimum threshold of 50,000 tons of CO_2 equivalent per year.

Demonstration of Additionality and Determination of Baseline Scenario and Emission Reductions

- Why should the project not happen on its own? (Does the project have significant barriers, or is it not the most economically attractive?)

- What would have happened in the absence of the project?
- What are the sources and total volume of emission reductions?

Competent Project Participants and Clear Institutional Arrangements

- The project must include technically experienced and sound project developers with a clear division of functions.
- The project must demonstrate sound legal arrangements—for example, who owns, who operates, and what type of agreement exists between project participants as well as with third parties (for example, power purchase agreement, ownership agreement, water rights).

Viable Business and Operation Model That Helps to Reduce Transaction Costs

- The project must demonstrate the potential for scale-up.
- The project must involve intermediaries who can invest, bundle, and implement project-related CDM services locally.

Ratification of Kyoto Protocol by the Host Country

- Has the host country ratified the Kyoto Protocol or expressed its intention to ratify the Kyoto Protocol in due course?
- What are the specific locations for implementation of the project?

Financing Sought

- The World Bank Carbon Finance Unit will not provide debt or equity finance for the baseline component of the project. The baseline component of the project should be financed by other sources.
- Payment will be made on delivery of emission reductions.

Sound Financing Structure

- The sound financial health of the project sponsors and co-financiers must be demonstrated.
- The sooner the project can achieve financial closure, the better the chances of selection.

Technical Summary of Project

- The project should be replicable or facilitate technology transfer for the country.
- The technology to be applied must be established and commercially feasible in a country other than the country in consideration.
- The project proposal should contain sample cases of the technology applied in the past in order to show its commercial feasibility.

Expected Environmental Benefits

- Evidence should be given that the project benefits are additional to the baseline or reference scenario, which represents the most likely or business-as-usual scenario in the country.

Safeguard Policies of the World Bank Group

- The Bank Group has a body of well-developed, mandatory safeguard policies that apply to all World Bank operations as well as an extensive set of good practices. These are applied to operations of the Carbon Finance Unit to ensure that they are environmentally and socially sound, whether baseline financing is from the Bank Group or from a third-party project supplier. The project must be consistent with these safeguard policies and the host country's overall sustainable development framework.

Contribution to Sustainable Development

- Contribution to sustainable development will be defined by the host country. For some end-of-pipe projects, contribution to sustainable development can be achieved by reinvesting some revenues from carbon finance in the host community.

Website http://go.worldbank.org/TV19LHL5J0

References

Aburto-Oropeza, O., E. Ezcurra, G. Danemann, V. Valdez, J. Murray, and E. Sala. 2008. "Mangroves in the Gulf of California Increase Fishery Yields." *Proceedings of the National Academy of Sciences* 105 (30): 10456–59.

Balmford, A., A. Bruner, P. Cooper, R. Constanza, S. Farber, R. Green, M. Jenkins, P. Jefferiss, V. Jessamy, J. Madden, K. Munro, N. Myers, S. Naeem, J. Paavola, M. Rayment, S. Rosendo, J. Roughgarden, K. Trumper, and R. T. Turner. 2002. "Economic Reasons for Conserving Wild Nature." *Science* 297 (5583): 950–53.

Barrett, K. R. 1999. "Ecological Engineering in Water Resources." *Water International* 24 (3): 182–88.

Baumert, K. A., T. Herzog, and J. Pershing. 2006. *Navigating the Numbers: Greenhouse Gas Data and International Climate Policy*. Washington, DC: World Resources Institute.

Bouillon, S., A. V. Borges, E. Castañeda-Moya, K. Diele, T. Dittmar, N. C. Duke, E. Kristensen, S. Y. Lee, C. Marchand, J. J. Middelburg, V. H. River-Monry, T. J. Smith III, and R. R. Twilley. 2008. "Mangrove Production and Carbon Sinks: A Revision of Global Budget Estimates." *Global Biogeochemical Cycles* 22: 1–12.

Caldeira, K., and M. E. Wickett. 2003. "Anthropogenic Carbon and Ocean pH." *Nature* 425 (6956): 365.

Campbell, A., V. Kapos, I. Lysenko, J. P. W. Scharlemann, B. Dickson, H. K. Gibbs, M. Hansen, and L. Miles. 2008. *Carbon Emissions from Forest Loss in Protected Areas*. Cambridge, U.K.: United Nations Environment Programme–World Conservation Monitoring Centre.

Corrigan, C., and F. Kershaw. 2008. *Working toward High Seas Marine Protected Areas: An Assessment of Progress Made and Recommendations for Collaboration*. Cambridge, U.K.: United Nations Environment Programme–World Conservation Monitoring Centre.

Damania, R., J. Seidensticker, A. Whitten, G. Sethi, K. MacKinnon, A. Kiss, and A. Kushlin. 2008. *A Future for Wild Tigers.* Washington, DC: World Bank.

Dudley, N., and S. Stolton, eds. 2003. *Running Pure: The Importance of Forest Protected Areas to Drinking Water.* Gland, Switzerland: World Wide Fund for Nature and World Bank Alliance for Forest Conservation and Sustainable Use.

Emerton, L., and E. Bos. 2004. *Value: Counting Ecosystems as an Economic Part of Water Infrastructure.* Water and Nature Initiative. Gland, Switzerland: World Conservation Union.

Emerton, L., N. Erdenesaikhan, B. de Veen, D. Tsogoo, L. Janchivdorj, G. Gavaa, Suvdaa, Enkhtsetseg, Ch. Dorjsuren, D. Sainbayar, and A. Enkhbaatar. 2009. *The Economic Value of the Upper Tuul Ecosystem, Mongolia.* Washington, DC: World Bank.

Falkowski, P., R. J. Scholes, E. Boyle, J. Canadell, D. Canfield, J. Elser, N. Gruber, K. Hibbard, P. Hoegberg, S. Linder, F. T. Mackenzie, B. Moore III, T. Pedersen, Y. Rosenthal, S. Seitzinger, V. Smetacek, and W. Steffen. 2000. "The Global Carbon Cycle: A Test of Our Knowledge of Earth as a System." *Science* 290: 291–96.

FAO (Food and Agriculture Organization). 2005. *Deforestation Continues at an Alarming Rate.* Rome, Italy: FAO.

GISP (Global Invasive Species Programme). 2008. *Assessing the Risk of Invasive Alien Species Promoted for Biofuels.* Nairobi, Kenya: GISP.

Gitay, H., A. Suárez, R. T. Watson, and D. J. Dokken, eds. 2002. *Climate Change and Biodiversity.* Technical Paper V. Geneva, Switzerland: IPCC.

Hirji, R., and R. Davis. 2009. *Environmental Flows in Water Resources Policies, Plans, and Projects: Case Studies.* Washington, DC: World Bank.

IFRC (International Federation of Red Cross and Red Crescent Societies). 2002. *Mangrove Planting Saves Lives and Money in Vietnam.* World Disasters Report Focus on Reducing Risk. Geneva, Switzerland: IFRC.

IPCC (Intergovernmental Panel on Climate Change). 2007. *Climate Change 2007: Synthesis Report.* Geneva, Switzerland: IPCC.

Janzen, D. 1999. "Gardenification of Tropical Conserved Wildlands: Multitasking, Multicropping, and Multiusers." *Proceedings of the National Academy of Sciences* 96 (11): 5987–94.

Kapos, V., C. Ravilious, A. Campbell, B. Dickson, H. Gibbs, M. Hansen, I. Lysenko, L. Miles, J. Price, J. P. W. Scharlemann, and K. Trumper, eds. 2008. *Carbon and Biodiversity: A Demonstration Atlas.* Cambridge, U.K.: United Nations Environment Programme–World Conservation Monitoring Centre.

Kramer, R. A., D. D. Richter, S. Pattanayak, and N. P. Sharma. 1997. "Ecological and Economic Analysis of Watershed Protection in Eastern Madagascar." *Journal of Environmental Management* 49 (33): 277–95.

Laffoley, D. d'A., ed. 2008. *Towards Networks of Marine Protected Areas: The MPA Plan of Action for IUCN's World Commission on Protected Areas.* Gland, Switzerland: IUCN, World Commission on Protected Areas.

Matthews, E., R. Payne, M. Rohweder, and S. Murray. 2000. *Pilot Analysis of Global Ecosystems: Forest Ecosystems.* Washington, DC: World Resources Institute.

Matthews, J., T. Le Quesne, R. Wilby, G. Pegram, J. Hartmann, B. Wickel, C. McSweeney, C. Von der Heyden, E. Levine, C. Guthrie, G. Blate, and G. de Freitas. 2009. *Flowing Forward: Making Decisions about Freshwater Biodiversity and Sustainable Management in a Shifting Climate.* Washington, DC: World Wildlife Fund–US.

Miles, L., and V. Kapos. 2008. "Reducing Greenhouse Gas Emissions from Deforestation and Forest Degradation: Global Land Use Implications." *Science* 320: 1454–55.

Moritz, C., J. L. Patton, C. J. Conroy, J. L. Parra, G. C. White, and S. R. Beissinger. 2008. "Impact of a Century of Climate Change on Small Mammal Communities in Yosemite National Park, USA." *Science* 322 (5899): 261–64.

Murphy, S. T., and O. D. Cheesman. 2006. *The Aid Trade: International Assistance Programs as Pathways for the Introduction of Invasive Alien Species; A Preliminary Report.* Washington, DC: World Bank.

Ong, J. E. 1993. "Mangroves: A Carbon Source and Sink." *Chemosphere* 27 (6): 1097–107.

Paris, R., and I. Ruzicka. 1991. *Barking up the Wrong Tree: The Role of Rent Appropriation in Sustainable Tropical Forest Management.* Occasional Paper 1. Philippines: Asian Development Bank.

Parish, F., A. Sirin, D. Charman, H. Joosten, T. Minayeva, M. Silvius, and L. Stringer, eds. 2008. *Assessment on Peatlands, Biodiversity, and Climate Change: Main Report.* Kuala Lumpur, Malaysia: Wetlands International.

Pena, N. 2009. *Including Peatlands in Post-2012 Climate Agreements: Options and Rationales.* Wageningen, the Netherlands: Wetlands International.

Peskett, L., D. Huberman, E. Bowen-Jones, G. Edwards, and J. Brown. 2008. *Making REDD Work for the Poor.* A Poverty Environment Partnership (PEP) Report. London, U.K.: Overseas Development Institute.

Pierce, S. M., R. M. Cowling, T. Sandwith, and K. MacKinnon, eds. 2002. *Mainstreaming Biodiversity in Development: Case Studies from South Africa.* Washington, DC: World Bank.

ProAct Network. 2008. *The Role of Environmental Management and Eco-engineering in Disaster Risk Reduction and Climate Change Adaptation.* Helsinki, Finland: Ministry of Environment; Geneva, Switzerland: Gaia Group and United Nations International Strategy for Disaster Reduction.

Quintero, J. D. 2007. *Mainstreaming Conservation in Infrastructure: Case Studies from Latin America.* Washington, DC: World Bank.

Ramsar Convention on Wetlands. 2005. *Background Papers on Wetland Values and Functions.* Ramsar.

Santili, M., P. Moutinho, S. Schwartzman, D. Nepstad, L. Curran, and C. Nobre. 2005. "Tropical Deforestation and the Kyoto Protocol: An Editorial Essay." *Climate Change* 71: 267–76.

Stern, N. 2007. *Stern Review of the Economics of Climate Change.* Cambridge, U.K.: Cambridge University Press.

Stickler, C., M. Coe, D. Nepstad, G. Fiske, and P. Lefebvre. 2007. *Ready for REDD? A Preliminary Assessment of Global Forested Land Suitability for Agriculture.* Woods Hole Research Center. Report released prior to the Bali UNFCCC meeting. http://whrc.org/policy/balireports.

Stolton, S., N. Dudley, and J. Randall. 2008. *Arguments for Protection: Natural Security, Protected Areas, and Hazard Mitigation.* Washington, DC: World Wildlife Fund.

Sukhdev, P. 2008. *The Economics of Ecosystems and Biodiversity.* Brussels: European Community.

van Zonneveld, M., J. Koskela, B. Vinceti, and A. Jarvis. 2009. "Impact of Climate Change on the Distribution of Tropical Pines in Southeast Asia." *Unasylva* 231 (60): 24–29.

Vergara, W. 2005. *Adapting to Climate Change: Lessons Learned, Work in Progress, and Proposed Next Steps for the World Bank in Latin America.* Washington, DC: World Bank.

Watson, R. T., I. R. Noble, B. Bolin, N. H. Ravindranath, D. J. Verardo, and D. J. Dokken, eds. 2000. *Land Use, Land Use Change, and Forestry.* Geneva, Switzerland: IPCC.

World Bank. 2008a. *Biodiversity, Climate Change, and Adaptation.* Washington, DC: World Bank.

———. 2008b. *World Development Report 2008: Agriculture for Development.* Washington, DC: World Bank.

———. Forthcoming. *Economics of Disaster Risk Reduction.* Washington, DC: World Bank.

WWF (World Wide Fund for Nature). 2008. *Water for Life: Lessons for Climate Change Adaptation from Better Management of Rivers for People and Nature.* Gland, Switzerland: WWF International.

INDEX

Boxes, figures, and tables are indicated by *b*, *f*, and *t*, respectively.

A

Aberdare Mountains National Park (Kenya), 85
Acacia senegalensis, 23–24, 25*b*
Aceh Forest and Environment Project (AFEP), Indonesia, 30–31*b*
ACG (Area de Conservación Guanacaste), Costa Rica, 71*b*
Ad Hoc Technical Expert Group (AHTEG) on Biodiversity and Climate Change, 90
adaptation, ecosystem-based, 3–4, 20, 49–65
 biodiversity conservation, 50–52, 51*b*
 defined, 49–50
 indigenous knowledge, making use of, 57, 58*b*
 infrastructure projects, integrating ecosystem protection and management into, 65, 68*t*
 investment in ecosystems, 63–65, 65*b*, 66–68*t*
 natural disasters, ecosystem management to prevent, 65, 66–67*t*
 restoration and maintenance of natural ecosystems, 41–42*b*, 52–53, 54*b*, 55*b*
 sustainable management projects, 53–55, 56*b*
Adaptation to the Impact of Rapid Glacier Retreat, Tropical Andes Project, 83–84
AFEP (Aceh Forest and Environment Project), Indonesia, 30–31*b*
Afghanistan, 44*t*
Africa. *See* Middle East and North Africa; sub-Saharan Africa; specific countries
African Development Bank, 95
agriculture, 4–5, 69–77
 agrobiodiversity, 74–75*b*
 biodiversity, as threat to, 69–70
 impact of climate change on, 4–5, 70–72, 74–75*b*

invasive alien species, threats posed by, 78–79
irrigation, 73*b*, 82–83
PES, 71*b*, 76, 78*b*
sustainable land management, 74–77, 76–77*b*
AHTEG (Ad Hoc Technical Expert Group) on Biodiversity and Climate Change, 90
Algeria, 16*t*
alien species. *See* invasive alien species
Aloe pillansii, 12*b*
alternative energy, 39–45
 biofuels, 42–45. *See also* biofuels
 hydropower, 39–42, 41–42*b*, 67*t*
Amazon ecosystem
 deforestation in, 36
 indigenous lands in, 57
 rainfall variability in, 17
 rising sea levels and, 15
 savanna likely to replace forest in, 15*b*
Amazon Region Protected Areas (ARPA) program, 37*b*
Analamazaotra Special Reserve (Madagascar), 29
Andhra Pradesh (India) water tank project, 73*b*
Ankeniheny-Mantadia-Zahamena Corridor Restoration and Conservation Carbon Project (Madagascar), 28–29
Arab Republic of Egypt, 16*t*
Area de Conservación Guanacaste (ACG), Costa Rica, 71*b*
Argentina
 biological corridors, 52
 in FCPF, 94
 flooding and flood control, 4, 15, 64, 65*b*
 invasive alien species in, 43*t*
ARPA (Amazon Region Protected Areas) program, 37*b*

105

Asia. *See* East Asia and Pacific; Eastern Europe and Central Asia; South Asia; specific countries
Atlantic Forest of Brazil, 51*b*
Australia, 35*f*, 43–44*t*, 84, 96*b*
Austria, 84
avalanches and landslides, 64–65, 67*t*

B

Bangladesh, 16*t*, 43*t*, 52, 92*t*
Benin, 16*t*
Berbak National Park (Indonesia), 32
Bhutan, 36, 52
BioCarbon Fund, 23*b*, 25*b*, 33*b*, 52, 90–91
biodiversity conservation, 4–6, 20, 69–80
  adaptive approach to, 50–52, 51*b*
  agriculture as threat to, 69–70
  agrobiodiversity, 74–75*b*
  coral reefs, 38
  forests, 30*b*
  grasslands, 33–34
  human communities and livelihoods, link to, 17–20, 69. *See also* agriculture
  impact of climate change on, 11–13, 12*b*, 13*b*
  PES, 71*b*, 76, 78*b*
  sustainable land management, 74–77, 76–77*b*
  wetlands, 32
biofuels, 42–45
  ethanol, 46–47*b*
  invasive alien species, 42, 43–44*t*
  oil palm plantations, 27*b*, 28, 29*b*, 31, 36, 43*t*, 45, 46*b*, 96*b*
  sustainability issues, 45, 46–47*b*
Biofuels Sustainability Scorecard, 45
biological corridors, 50–52, 51*b*
biological mitigation. *See* mitigation, ecosystem-based
Bokkeveld plateau (South Africa), 76–77*b*
Bolivia
  biological corridors, 52
  in FCPF, 94
  forest area and forest carbon stocks, 27*f*
  human benefits from ecosystem protection in, 92*t*
  protected areas in, 36
  UN-REDD, 95
  water supply, 83
Brazil. *See also* Amazon ecosystem
  Atlantic Forest biological corridor, 51*b*
  biofuel production in, 43–44*t*, 45, 46*b*
  climate threats to, 15
  forests, 27*b*
  indigenous territories in, 57
  invasive alien species in, 43–44*t*
  protected areas in, 36, 37*b*
  reforestation of hydroreservoirs, 23*b*
  urban sources of water in, 84
Bulgaria, 36, 53, 54*b*
Burundi, 75

C

Cambodia, 16*t*, 92*t*
Cameroon, 94
CAPE Biodiversity Conservation and Sustainable Development Project, 80
Cape Floristic Region (CFR), South Africa, 11, 79–80
carbon emissions. *See* greenhouse gases
Caribbean. *See* Latin America and the Caribbean
CARICOM (Caribbean Community) countries, impact of climate change on, 59
Cayos Miskitos Biological Reserve (Caribbean coast), 51*b*
CBD (Convention on Biological Diversity), 18, 61, 90
CDM (Clean Development Mechanism), 22, 90–91, 92, 97, 98
Central America. *See* Latin America and the Caribbean
Central Asia. *See* Eastern Europe and Central Asia
CEPF (Critical Ecosystem Partnership Fund), 51*b*
Cerro Silva Natural Reserve (Caribbean coast), 51*b*
CFR (Cape Floristic Region), South Africa, 11, 79–80
Chaco Andean system, 51*b*
Chad, 16*t*
Chile, 11, 43*t*
China
  climate threats to, 16*t*
  Forest Protection Project, 53–54
  Grain for Green Program, 67*t*
  grasslands conservation, 34*b*
  Hövsgöl National Park research in Mongolia and, 13*b*
  invasive alien species in, 44*t*
  reforestation of Pearl River watershed, 23*b*
  urban sources of water in, 85
  wastewater treatment in, 82–83*b*
  water supply, 6

CIFs (Climate Investment Funds), 91
Clean Development Mechanism (CDM), 22, 90–91, 92, 97, 98
Clean Technology Fund, 91
climate change, ecosystem-based approaches to, 1–7, 9–20
  adaptation, 3–4, 20, 49–65. *See also* adaptation, ecosystem-based
  biodiversity, conserving, 4–6, 20, 69–80. *See also* biodiversity
  carbon cycle, role of terrestrial and marine ecosystems in, 10–11, 10*f*
  GHGs and climate change. *See* greenhouse gases
  impact of climate change
    on agriculture, 4–5, 70–72, 74–75*b*. *See also* agriculture
    on ecosystems and biodiversity, 11–13, 12*b*, 13*b*
    on human communities and livelihoods, 14–15*b*, 14–17, 16*t*
  implementation of, 6–7, 20, 87–95. *See also* implementation of ecosystem-based approaches
  importance of addressing, 17–20
  mitigation, 2–3, 20, 21–45. *See also* mitigation, ecosystem-based
  World Bank and, 1–2, 9–10, 18–20, 19*t*, 87–88, 97–99. *See also* BioCarbon Fund; Global Environment Facility; implementation of ecosystem-based approaches; specific projects and areas
Climate Investment Funds (CIFs), 91
coastal wetlands and mangrove swamps
  adaptive approach to, 57–61, 59*b*, 60*b*
  biological mitigation via, 32–33
  natural disasters, ecosystem management to prevent, 66*t*
Colombia
  biological corridors, 50–52, 51*b*
  in FCPF, 95
  flooding and flood control, 15
  forests, 27*b*
  indigenous territories in, 57
  PES, 78*b*
  protected areas in, 36
  urban sources of water in, 84
Congo Basin Forest Fund, 95
Congo, Democratic Republic of, 27*b*, 94, 95
Congo, Republic of, 94
Convention on Biological Diversity (CBD), 18, 61, 90
Coral Reef Rehabilitation and Management Project (COREMAP), 62–63*b*

Coral Reef Targeted Research and Capacity Building for Management (CRTR) Program, 60*b*
coral reefs
  biodiversity of, 38
  biological mitigation and, 22, 38–39, 39*b*
  as carbon sinks, 28–29
  climate change, vulnerability to, 11, 13, 14, 16, 21, 22*f*
  CPACC on, 59–60
  economics of protecting, 39*b*
  International Year of the Reef (2008), 60*b*
  Manado Ocean Declaration, 39, 40*b*
  natural disasters, ecosystem management to prevent, 66*t*
  protected areas, 61, 62–63*b*, 64*b*
  restoration and maintenance efforts, 52
  World Bank funding to conserve, 18
Coral Triangle Initiative, 63, 64*b*
COREMAP (Coral Reef Rehabilitation and Management Project), 62–63*b*
Costa Rica, 55, 56*b*, 71*b*, 78*b*, 95
CPACC (Caribbean Planning for Adaptation to Climate Change), 58–60
Critical Ecosystem Partnership Fund (CEPF), 51*b*
Croatia, 36, 52
CRTR (Coral Reef Targeted Research and Capacity Building for Management) Program, 60*b*
Czech Republic, 64

D

Danube River wetlands, restoration and maintenance of, 53, 54*b*
degradation and desertification, sustainable land management to avoid, 74–77, 76–77*b*
Del Oro Company, 71*b*
Democratic Republic of Congo, 27*b*, 94, 95
desertification and degradation, sustainable land management to avoid, 74–77, 76–77*b*
Djibouti, 44*t*
drought, 5, 70–72, 75
Dumoga-Bone National Park (Indonesia), 52

E

East Asia and Pacific. *See also* specific countries
  Coral Triangle Initiative, 63, 64*b*
  invasive alien species in, 43–44*t*
  protected areas in, 35*f*
  regional impact of climate change on, 14–15*b*

Eastern Europe and Central Asia. *See also* specific countries
  agricultural activity, climate change, and drought in, 5, 70–71
  biological corridors, 52
  Hövsgöl National Park research in Mongolia and, 13*b*
  protected areas in, 35*f*, 36
  regional impact of climate change on, 14*b*
EBAs (Endemic Bird Areas), 18, 33
Ecomarkets Project, 55, 56*b*
economics of ecosystem-based approaches. *See* financing ecosystem-based approaches
ecosystem-based approaches to climate change. *See* climate change, ecosystem-based approaches to
Ecuador
  Adaptation to the Impact of Rapid Glacier Retreat, Tropical Andes Project, 83–84
  biological corridors, 51*b*, 52
  flooding and flood control, 4, 64
  invasive alien species in, 43*t*
  protected areas, funding for, 36
  sustainable management of ecosystems in, 55
  water supply, 6, 84, 85
Egypt, Arab Republic of, 16*t*
El Triunfo Reserve (Mexico), 54
elephants, 30*b*, 41*b*, 50, 96*b*
Endemic Bird Areas (EBAs), 18, 33
Eritrea, 16*t*, 44*t*
ethanol, 46–47*b*
Ethiopia, 16*t*, 44*t*, 74, 94
Europe. *See* Eastern Europe and Central Asia; Western Europe; specific countries

F

farming. *See* agriculture
FCPF (Forest Carbon Partnership Facility), 7, 90, 93–95
Fiji, 15, 16*t*, 43*t*
financing ecosystem-based approaches
  CIFs, 91
  developing financing mechanisms, 90–91
  minimum project requirements for securing World Bank carbon finance, 97–99
  total and World Bank Group funding statistics, 18, 19*t*
FIP (Forest Investment Program), 7, 91, 95
fire risk posed by invasive alien species, 80
flooding and flood control, 4, 15, 64, 65*b*, 66*t*, 81*b*, 82
flower farming, sustainable, 76–77*b*
Fondo Nacional de Financiamiento Forestal (FONAFIFO), Costa Rica, 56*b*
Forest Carbon Partnership Facility (FCPF), 7, 90, 93–95
Forest Investment Program (FIP), 7, 91, 95
forests
  afforestation and reforestation, 22–26, 23–25*b*
  biological mitigation via, 2, 22–29
  carbon stores in, 26*t*
  deforestation and degradation, consequences of, 26–28, 27*f*
  as flood control and water conservation method, 41–42*b*, 64, 65*b*, 66*t*, 81*b*
  GFP initiative, 89–90
  hydropower and, 41–42*b*, 67*t*
  landslides and avalanches, as protection against, 64–65, 67*t*
  multiple downstream benefits of conserving, 81*b*
  REDD, 2, 28–29, 31*b*, 91–95, 94*b*
  silvopastoral systems, 34–35, 76
  sustainable management of, 28–30, 29*b*, 30–31*b*, 53–54
funding. *See* financing ecosystem-based approaches
fynbos, 11

G

Gabon, 36, 94
Galapagos Islands, 79
Gansu and Xinjiang Pastoral Development Project (China), 34*b*
GEF. *See* Global Environment Facility
Georgia (country), 36
Germany, 56*b*, 95
GFP (Growing Forest Partnerships) initiative, 89–90
Ghana, 36, 94
GHGs. *See* greenhouse gases
GISP (Global Invasive Species Programme), 79
Glen Lyon (South Africa), 76–77*b*
Global Environment Facility (GEF)
  Adaptation Fund, 91
  agricultural projects, 74*b*, 76*b*, 78*b*, 80
  biodiversity investments by, 19*t*, 50–52, 51*b*, 74*b*, 78*b*, 80
  carbon sequestration and storage, standard measures of, 92
  Hövsgöl National Park, Mongolia, grant for, 13*b*
  indigenous knowledge, making use of, 58*b*
  protected areas, funding for, 36, 37*b*

Richtersveld landscape, funding for, 12*b*
sustainable management projects, 55, 56*b*
wetlands conservation in Indonesia, 32
Global Facility for Disaster Reduction and
 Recovery, 91
Global Invasive Species Programme
 (GISP), 79
global warming. *See* climate change,
 ecosystem-based approaches to
grasslands
 biodiversity and, 33–34
 biological mitigation via, 3, 33–35
 carbon stores in, 26*t*, 33
 conservation and management,
  34–35, 34*b*
 silvopastoral systems, 34–35, 76
greenhouse gases (GHGs)
 atmospheric increase in, 21
 biofuels and, 45
 carbon markets, 91, 95, 96*b*
 climate change triggered by, 1, 9,
  21–22, 22*f*
 ecosystem-based reduction of, 2–3, 11
 ecosystems as carbon stores and sinks. *See*
  mitigation, ecosystem-based
 emphasis on reducing, 2, 17
 REDD, 2, 28–29, 31*b*, 91–95, 94*b*
 standard measures of carbon sequestration
  and storage, 92
 Sumatran tigers and elephants, carbon
  trading as means of saving, 96*b*
 UN-REDD, 95
Growing Forest Partnerships (GFP) initiative,
 89–90
Guatemala, 51*b*
Gunung Gede Pangrango National Park
 (Indonesia), 6, 85
Gunung Leuser National Park
 (Indonesia), 30*b*
Guyana, 95

H

Haiti, 16*t*, 66*t*, 85
Honduras, 16*t*, 51*b*, 52
Hövsgöl National Park, Mongolia, 13*b*
human communities and livelihoods
 biodiversity, link to, 17–20
 ecosystem protection, potential benefits
  from, 91, 92*t*
 impact of climate change on, 14–15*b*,
  14–17, 16*t*
 sustainable management projects
  contributing to, 53–55, 56*b*
hydropower, 39–42, 41–42*b*, 67*t*

I

IBAs (Important Bird Areas), 18, 24
IBRD (International Bank for Reconstruction
 and Development), 19*t*
IDA (International Development
 Association), 19*t*
implementation of ecosystem-based
 approaches, 6–7, 20, 87–95
 CIFs, 91
 FCPF, 7, 90, 93–95
 financing. *See* financing ecosystem-based
  approaches
 GFP initiative, 89–90
 REDD, 2, 28–29, 31*b*, 91–95, 94*b*
 SFCCD, 87, 88–89
 UN-REDD, 95
Important Bird Areas (IBAs), 18, 24
India
 agricultural activities in, 5, 69, 73*b*, 75
 biological corridors, 51*b*
 climate threats to, 16*t*
 invasive alien species in, 43–44*t*, 79
 protected areas in, 36
 urban sources of water in, 84
 water tank irrigation in, 73*b*
indigenous knowledge, making use of,
 57, 58*b*
Indonesia
 biofuel production in, 44*t*, 45
 biological corridors, 51*b*
 climate threats to, 16*t*
 Coral Triangle Initiative, 64*b*
 COREMAP, 62–63*b*
 forests of, 27*b*, 28, 30–31*b*
 invasive alien species in, 44*t*
 protected areas in, 36
 restoration and maintenance of natural
  ecosystems in, 52
 Sumatran tigers and elephants, carbon
  trading as means of saving, 96*b*
 tsunami in, 30–31*b*
 UN-REDD, 95
 water supply, 6, 84, 85
 wetlands and peatlands in, 31–32, 45
infrastructure projects, integrating ecosystem
 protection and management into,
 65, 68*t*
Intergovernmental Panel on Climate Change
 (IPCC), 14–15*b*, 21, 92
International Bank for Reconstruction
 and Development (IBRD), 19*t*
International Development Association
 (IDA), 19*t*

International Union for Conservation of
    Nature (IUCN), 36
International Year of the Reef (2008), 60b
invasive alien species, 5
    as biodiversity and agricultural threat,
        77–80, 80b
    biofuel production and, 42, 43–44t
IPCC (Intergovernmental Panel on Climate
    Change), 14–15b, 21, 92
Iran, Islamic Republic of, 16t
irrigation, 73b, 82–83
Islamic Republic of Iran, 16t
itch grass, 5, 79
IUCN (International Union for Conservation
    of Nature), 36

J

Japan, 43t, 56b, 66t, 84
Jatropha curcas, 45
Jeanne (hurricane), 66t

K

Katrina (hurricane), 66t
Kazakhstan, 52
Kenya
    biofuel production in, 45
    carbon payments pilot program, 91
    climate threats to, 16t
    in FCPF, 94
    reforestation in, 24b
    water supply, 6, 84, 85
Kerinci-Seblat National Park (Indonesia), 6, 85
KfW (Kreditanstalt für Wiederaufbau), 37b
Kiribas, 58
Kiribati, 43t
Kreditanstalt für Wiederaufbau (KfW), 37b
Kyoto Protocol
    CDM, 22, 90–91, 92, 97, 98
    forest management and emission reduction
        targets, 28, 29
    Sumatran tigers and elephants, carbon
        trading as means of saving, 96b
    World Bank Carbon Finance Unit
        minimum project requirements,
        97, 98
Kyrgyz Republic, 52

L

land degradation and desertification,
    sustainable management to avoid,
    74–77, 76–77b
Land Use, Land Use Change, and Forestry
    (LULUCF) projects, 23b

landslides and avalanches, 64–65, 67t
Lao People's Democratic Republic (Lao PDR)
    climate threats to, 16t
    in FCPF, 95
    Nam Theun 2 Hydropower Project, 40,
        41–42b, 52
Latin America and the Caribbean. *See also*
    specific countries
    biofuel production in, 45
    CARICOM countries, impact of climate
        change on, 59
    CPACC on climate change impacts, 58–60
    FCPF, countries in, 94–95
    invasive alien species in, 5, 43–44t, 79
    protected areas, 35f, 36
    regional impact of climate change on, 14,
        15–16, 15b
    silvopastoral projects in, 34–35
    sustainable land management in, 76–77
Lesotho, 42, 52
Liberia, 94
Libya, 16t
Line Islands, 43t
Lithuania, 52
Lower Danube Green Corridor, 53, 54b
LULUCF (Land Use, Land Use Change, and
    Forestry) projects, 23b
Lužnice floodplain (Czech Republic), 64

M

MABC (Mesoamerican Biological Corridor),
    51b, 56b
Madagascar
    biological corridors, 51b
    climate threats to, 16t
    in FCPF, 94
    flooding and flood control in, 66t
    forest preservation in, 28–29, 66t, 81b
    invasive alien species in, 43t
    natural disasters, ecosystem-based
        prevention of, 66b
    protected areas in, 36
    sustainable land management in, 74
maintenance and restoration of natural
    ecosystems, 41–42b, 52–53, 54b, 55b
Malawi, 16t
Malaysia, 27b, 44t, 59b, 64b
Mali, 16t, 24, 25b, 42
Maloti-Drakensberg transfrontier region
    (southern Africa), 52
Manado Ocean Declaration, 39, 40b
mangrove swamps. *See* coastal wetlands and
    mangrove swamps
Mantadia National Park (Madagascar), 29, 66t

marine ecosystems. *See* oceans
Marrakesh Accords, 97
Mauritania, 16*t*, 42, 74
Mauritius, 36, 43*t*
McGregor, Neil, 76*b*
MDGs (Millennium Development Goals), 10, 69
Mesoamerican Barrier Reef Project, 62
Mesoamerican Biological Corridor (MABC), 51*b*, 56*b*
Mexican Nature Conservation Fund, 54
Mexico
 biological corridors, 51*b*
 climate threats to, 16*t*
 in FCPF, 95
 invasive alien species in, 43–44*t*
 sustainable management of ecosystems in, 54–55
 water supply, 6, 84, 85
 wetlands preservation and management, 55*b*
Micronesia, Federated States of, 43*t*
Middle East and North Africa. *See also* specific countries
 agricultural activity, climate change, and drought in, 5, 70–71
 protected areas in, 35*f*
 sustainable management of ecosystems in, 53
Millennium Development Goals (MDGs), 10, 69
Millennium Ecosystem Assessment, 10, 70
mitigation, ecosystem-based, 2–3, 20, 21–45
 alternative energy, investing in, 39–45. *See also* alternative energy
 defined, 22
 different terrestrial and marine ecosystems, roles of, 2–3, 10–11, 10*f*, 26–35
 forests, 2, 22–29. *See also* forests
 GHG rise, effects of not mitigating, 21–22, 22*f*
 grasslands, 3, 33–35. *See also* grasslands
 oceans, 3, 38–39, 38–40*b*. *See also* coastal wetlands and mangrove swamps; coral reefs; oceans
 protected areas, 35–36, 35*f*, 37*b*
 wetlands, 2–3, 30–33. *See also* coastal wetlands and mangrove swamps; wetlands
Moldova, 16*t*, 54*b*
Monarch Butterfly Reserve (Mexico), 54, 85
Mongolia, 13, 16*t*, 85, 86*b*
Morocco, 16*t*, 53
Mount Kenya National Park (Kenya), 85
mountain ecosystems, 16–17, 53

Mozambique, 16*t*, 92*t*
Muthurajawela marsh (Sri Lanka), 82
Myanmar, 16*t*, 43*t*, 57

N

Nakai Nam Theun (Lao PDR), 41–42*b*
Nam Theun 2 Hydropower Project (Lao PDR), 40, 41–42*b*, 52
Nama Karoo (South Africa), 77*b*
Namibia, Succulent Karoo biome in, 12*b*
Namibian Coast Conservation and Management Project, 62
Nargis (cyclone; 2008), 57
Nariva wetlands restoration project (Trinidad and Tobago), 32–33*b*
native vegetation, promoting, 25*b*
natural disasters. *See also* specific disasters, e.g. Katrina
 climate change worsening, 16*t*
 ecosystem management to prevent, 65, 66–67*t*
Nepal, 92*t*, 95
Netherlands, 96*b*
New Caledonia, 43*t*
New Zealand, 35*f*, 43–44*t*
Nicaragua, 51*b*, 57, 78*b*, 95
Niger, 16*t*, 24, 92*t*
North Africa. *See* Middle East and North Africa, and specific countries
North America. *See also* Mexico; United States
 invasive alien species in, 43*t*
 protected areas in, 35*f*
Norway, 56*b*, 95
Nusa Tenggara, 62*b*

O

oceans. *See also* coastal wetlands and mangrove swamps; coral reefs
 biological mitigation via, 3, 38–39, 38–40*b*
 climate change affecting, 14, 38*b*
 Manado Ocean Declaration, 39, 40*b*
 overfishing, 61
 protected areas, 61–63, 62–63*b*, 64*b*
 rising sea levels, 14–15
oil palm plantations, 27*b*, 28, 29*b*, 31, 36, 43*t*, 45, 46*b*, 96*b*
overfishing, 61

P

Pacific Islands. *See* East Asia and Pacific
Padeido Islands, 62*b*
Pakistan, 16*t*, 36, 84

palm oil production, 27*b*, 28, 29*b*, 31, 36, 43*t*, 45, 46*b*, 96*b*
Panama, 51*b*, 95
panda reserves, Qinling Mountains (China), 85
Papua New Guinea, 36, 62*b*, 64*b*, 95
Paraguay, 95
Participatory Management of Protected Areas Project (Peru), 58*b*
payments for ecosystem services (PES), 54–55, 56*b*, 71*b*, 76, 78*b*
peatlands, 2–3, 30–31, 45
Peru
    Adaptation to the Impact of Rapid Glacier Retreat, Tropical Andes Project, 83–84
    biological corridors, 52
    in FCPF, 95
    forests, 27*b*
    protected areas in, 36
    Salinas and Aguada Blanca National Reserve, management of, 58*b*
PES (payments for ecosystem services), 54–55, 56*b*, 71*b*, 76, 78*b*
Philippines, 15, 16*t*, 51*b*, 64*b*
Pic Macaya National Park (Haiti), 85
Pilot Program for Climate Resilience (PPCR), 7, 91
Pilot Program to Conserve the Brazilian Rain Forest, 51*b*
PPCR (Pilot Program for Climate Resilience), 7, 91
protected areas
    as carbon sinks, 35–36, 35*f*, 37*b*
    marine reserves, 61–63, 62–63*b*, 64*b*

Q

Qinling Mountains panda reserves (China), 85

R

Ramsar Convention on Wetlands (2005), 59*b*
Ramsar sites, 32, 32*b*
Reducing emissions from deforestation and forest degradation (REDD), 2, 28–29, 31*b*, 91–95, 94*b*
Regional Integrated Silvopastoral Approaches to Ecosystem Management, 78*b*
Republic of Congo, 94
restoration and maintenance of natural ecosystems, 41–42*b*, 52–53, 54*b*, 55*b*
Richtersveld cultural and botanical landscape, 12*b*
Romania, 54*b*
Russian Federation, 13*b*, 28, 36
Rwanda, 16*t*

S

Salinas and Aguada Blanca National Reserve (Peru), 58*b*
salt marshes. *See* coastal wetlands and mangrove swamps
Samoa, 16*t*, 36
SCF (Strategic Climate Fund), 91, 95
seas. *See* oceans
Sembilang National Park (Indonesia), 32
Senegal, 15, 16*t*, 40–42, 74
Serengeti ecosystem, 34
Seychelles, 36, 66, 79
SFCCD (Strategic Framework for Climate Change and Development), 87, 88–89
silvopastoral systems, 34–35, 76
Singapore, 84
Solomon Islands, 64*b*
South Africa
    biological corridors in, 52
    CFR, 11, 79–80
    conservation farming in, 76–77*b*
    invasive alien species in, 43–44*t*, 79–80, 80*b*
    protected areas in, 36
    Succulent Karoo biome in, 12*b*
    urban sources of water in, 84
South America. *See* Latin America and the Caribbean
South Asia. *See also* specific countries
    agricultural activity and climate change in, 5, 72
    biological corridors, 51*b*
    irrigation in, 82–83
    protected areas in, 35*f*
    regional impact of climate change on, 14–15*b*
    water supply, 5
    wetlands conversion and conservation in, 31–33
Spain, 84
Sri Lanka, 16*t*, 43*t*, 66*t*, 82
Strategic Climate Fund (SCF), 91, 95
Strategic Framework for Climate Change and Development (SFCCD), 87, 88–89
sub-Saharan Africa. *See also* specific countries
    agricultural activity and climate change in, 5, 72, 74
    biofuel production in, 45
    biological corridors, 51*b*
    FCPF, countries in, 94
    grasslands and biodiversity, 34
    invasive alien species in, 43–44*t*, 79
    protected areas in, 35*f*

regional impact of climate change on, 14*b*
Succulent Karoo biome, 12*b*, 76*b*
Succulent Karoo biome, 12*b*, 76*b*
Sudan, 16*t*, 44*t*
Switzerland, 65, 67*t*

T

Tajikistan, 92*t*
Taka Bone Rate (Indonesia), 62*b*
Tanzania, 36, 62, 84, 85, 95
TerrAfrica, 74
Thailand, 16*t*, 43*t*, 44*t*, 59*b*
Tien Shan ecosystem, 52
tigers, 28, 30*b*, 32, 96*b*
Timor Leste, 64*b*
Tonga, 16*t*
Trinidad and Tobago, 32–33*b*
Tropical Andes Project, 83–84
tsunami disaster (2004), 30–31*b*, 57, 66*t*
Tunisia, 16*t*
Turkey, 53

U

Uganda, 36, 94
Ukraine, 54*b*
Ulu Masen Forest Complex (Indonesia), 30*b*, 31*b*
UN-REDD (United Nations Collaborative Program on Reducing Emissions from Deforestation and Forest Degradation in Developing Countries), 95
UNDP (United Nations Environment Programme), 92
UNESCO (United Nations Educational, Scientific, and Cultural Organization), 12*b*
UNFCCC (United Nations Framework Convention on Climate Change), 22, 28, 40*b*, 90–91, 95
United Kingdom, 61, 66*t*
United Nations
  Collaborative Program on Reducing Emissions from Deforestation and Forest Degradation in Developing Countries (UN-REDD), 95
  Educational, Scientific, and Cultural Organization (UNESCO), 12*b*
  Environment Programme (UNDP), 92
  Framework Convention on Climate Change (UNFCCC), 22, 28, 40*b*, 90–91, 95
United States
  flooding and flood control in, 66*t*
  grasslands and biodiversity, 33–34
  invasive alien species in, 43–44*t*, 79
  urban sources of water in, 84
urban sources of water, 84–85, 86*b*

V

Vanuatu, 95
Venezuela, República Bolivariana de, 15, 27*f*, 52
Vietnam
  climate threats to, 16*t*
  in FCPF, 95
  forest conservation on Lao border, 41*b*
  mangrove swamps in, 58, 59*b*, 62
  protected areas in, 36
  UN-REDD, 95
Vilicabamba-Amboró region (Latin America), 52
La Visite National Park (Haiti), 85

W

wastewater treatment, 82–83*b*, 84, 85*b*
water supply
  climate threats to, 4–6, 16
  drought, 5, 70–72, 75
  flooding and flood control, 4, 15, 64, 65*b*, 66*t*, 81*b*, 82
  forests, role of, 41–42*b*, 64, 65*b*, 66*t*, 67*t*, 81*b*
  hydropower, 39–42, 41–42*b*, 67*t*
  infrastructure projects, integrating ecosystem protection and management into, 65, 68*t*
  irrigation, 73*b*, 82–83
  natural ecosystems, protecting, 81–84, 81*b*, 82–83*b*
  natural water towers, 84–85, 86*b*
  Salinas and Aguada Blanca National Reserve (Peru), 58*b*
  urban sources, 84–85, 86*b*
  wastewater treatment, 82–83*b*, 84, 85*b*
  wetlands, role of, 82–83*b*, 84, 85*b*
Wawashan Natural Reserve (Caribbean coast), 51*b*
Western Europe
  invasive alien species in, 43–44*t*
  protected areas in, 35*f*
wetlands. *See also* coastal wetlands and mangrove swamps
  biodiversity, 32
  biological mitigation via, 2–3, 30–33
  carbon stores in, 26*t*
  conservation and management of, 31–33

drainage of, 31
flooding and flood control, 66*t*, 82
peatlands, 2–3, 30–31, 45
restoration and maintenance efforts, 53, 54*b*, 55*b*
sustainable management of, 53
water supply and, 82–83*b*, 84, 85*b*
Wetlands International, 31–32
World Bank and climate change, 1–2, 9–10, 18–20, 19*t*, 87–88, 97–99. *See also* BioCarbon Fund; Global Environment Facility; implementation of ecosystem-based approaches; specific projects and areas
World Heritage List, 12*b*

World Wide Fund for Nature (WWF), 18, 37*b*, 53
World Wildlife Fund, 45
Wukong (typhoon; 2000), 59*b*
WWF (World Wide Fund for Nature), 18, 37*b*, 53

Y

Yemen, Republic of, 53, 74–75*b*

Z

Zambia, 16*t*, 92*t*, 95
Zimbabwe, 16*t*

ECO-AUDIT
Environmental Benefits Statement

The World Bank is committed to preserving endangered forests and natural resources. The Office of the Publisher has chosen to print ***Convenient Solutions to an Inconvenient Truth*** on recycled paper with 100 percent postconsumer fiber in accordance with the recommended standards for paper usage set by the Green Press Initiative, a nonprofit program supporting publishers in using fiber that is not sourced from endangered forests. For more information, visit www.greenpressinitiative.org.

Saved:
- 19 trees
- 6 million British thermal units of total energy
- 1,845 pounds of net greenhouse gases
- 8,885 gallons of waste water
- 539 pounds of solid waste